THE Newfoundland COYOTE

by

Darrin McGrath

DRC Publishing
3 Parliament Street
St. John's, Newfoundland
A1A 2Y6

(709) 726-0960
e-mail: staceypj@avint.net

Library and Archives Canada Cataloguing in Publication

McGrath, Darrin Michael, 1966 -
The Newfoundland Coyote / Darrin McGrath. -- 1st ed.

Includes index
ISBN number 0-9684209-6-6

1. Coyote--Newfoundland. I. Title.

QL737.C22M34 2004   599.77'25'09718   C2004-906810-5

Cover Design by Diane Lynch

Front cover photo by Dale Wilson Photography

Back cover photo appears courtesy Brent Sellars, Newfoundland and Labrador Hydro Company

Printed in St. John's, Newfoundland, by Hutton International Press

This book is dedicated to my sister Rosemarie (February 9, 1952-March 25, 2004). Rosemarie was always ready to help proofread my written work and she loved dogs.

Table of Contents

Acknowledgments

I am an avid hunter and fishermen and I have followed the trail of the coyote in Newfoundland with much interest. As a free-lance writer I have had occasion to write stories about the "little wolf" for newspapers and magazines such as Outdoor Canada. My interest in the animal reached a point where I felt compelled to write a book about the coyote in Newfoundland.

The primary source of information for the book was interviews with about two dozen subjects from various fields such as agriculture, wildlife biology and management, outfitting and trapping, and resident hunters. I owe a debt of thanks to many people.

Several people loaned me photos: George Jennings, John Blake, John Neville, Paul Dunphy, Brent Sellars, Kelley Symonds, Gary Tuff and Randy Dibblee. John Blake also proofread for me. Thanks to Mike McGrath for showing me coyote carcasses in the wildlife lab.

Jonathan Crowe of CBC TV unselfishly shared his list of contacts stemming from his groundbreaking documentary about the coyote. Jonathan was a great help in getting me started.

Linda Swain of VOCM Radio was also very helpful in facilitating the collection of coyote photographs. Her interest in the project is greatly appreciated.

A sincere thanks to the following people who answered my many questions and shared their stories with me: Deborah Guillemette, Executive Director of the Newfoundland Federation of Agriculture; Dr. Hugh Whitney; Kim Bridger; Bill Green; Eric Patey; Wayne Barney; Eugene Tiller; Monroe Greening; Wayne Holloway; Rick Bouzan; Gordon Cooper; Shane Mahoney; Howard Morry; Wade Alley; Norma Collet; Gerald Hewitt; Pat Hewitt; Neil Coffey; Bruce Porter; Chad Snow; Jean Kennedy; Cyril Myrick; Gord Follet; Ed McGrath; Ivan Pitcher; Cliff Head; Dan Hutton of Hutton International Press. Thanks to Diane Lynch for excellent design and layout.

Special thanks to: Barry Sabean, Director of Wildlife Management for Nova Scotia; Cade Libby, Fur-bearer Management Biologist, Dept. Natural Resources, New Brunswick; Randy Dibblee, Widlife Biologist, Prince Edward Island Fish and Wildlife Division.

Hats off to Jim Wellman, Managing Editor of The Navigator magazine, which I write for every month.

A special thanks to my wife Ann for her unwavering encouragement, love, support and proof-reading.

Thanks to my mother Mary who is always interested in my projects and who bought me a wildlife encyclopedia when I was a child.

A sincere thanks to my brothers Pat and Jim who are always nearby when needed. My sister Marilyn can always be counted on to offer encouragement when the load gets heavy. As usual she played a vital role in the creation of this book.

Likewise, my mother and father-in-law, Dave and Mary Lou Stinson are always right behind me.

I owe a debt of gratitude to thank Peter and Jean Stacey for coming on-board at a crucial phase of this project. They helped ensure that this book was created.

Errors or omissions are my responsibility.
Darrin McGrath, October 2004

PREFACE

It was mid-March 1985 and the Gulf of St. Lawrence was blocked with heavy sea-ice.

As night's inky blackness gave way to the half-light of a grey dawn, the three travelers stopped to rest and take their bearings.

It was going to be a fine morning, the leader sensed. It had been two nights since the scent of seal blood had drawn them onto the ice and their bellies were beginning to ache with hunger. Now the wind baffled around from the east, carrying with it the unmistakable smell of spruce trees and caribou.

The heavy scent of hot-blooded ungulates caught their attention, making the hair stand up on their necks in nervous anticipation. The leader licked his lips and barked a marching order. They quickly broke into a loping trot across the open ice field.

As the sun began to rise they were a mile or two from shore.

...CHAPTER ONE...

Arrival

It was just before St. Patrick's Day, 1985 and veteran Wildlife Officer Bill Green had just arrived in his Pasadena office on Newfoundland's west coast when the phone started ringing. Bill hadn't even had time to grab a coffee and he wondered what was going on now. He wondered if it was a complaint about a poached moose, because there seemed to have been a slight increase in poaching this winter. Much to his surprise the caller was fellow Wildlife Officer Pete St. Croix, stationed on the Port au Port Peninsula.

"Bill, we've got a report of three wolves on the ice near Marches Point here on the Port au Port," St. Croix said.

Bill remembers the morning well. "It was just before St. Patrick's Day 1985 and the ice conditions were severe. The Gulf was blocked. As soon as we got the report from Pete St. Croix we took the chopper and went out to the approximate site."

Green says that it was a crystal clear morning and the chopper scoured the ice-fields off the Port au Port Peninsula for about 15 kilometers out to sea. He recollects that the chopper probably spent about an hour in the air that morning. The pilot and Green figured that on such a clear morning they would have no trouble spotting three darkish colored wolves or dogs. The men also flew over

1

several groups of caribou that inhabited the Port au Port at that time, but to no avail.

"There were caribou on the Port au Port then, they had been put there in the 1960s. We figured if these wolves or dogs had got ashore they'd go after the caribou. We checked every clump of caribou we knew of but we couldn't see any tracks. And there was no report of sheep killed either," Green says.

The wildlife officer heard nothing more about the three wolf-like animals until about a year later trapper George Jennings of Hughes Brook, on the north shore of the Bay of Islands, came into possession of a funny looking fox.

Jennings has been trapping for forty years and he knows the woods that surround Corner Brook and Deer Lake like the back of his hand. He also knows fur-bearers and remembers the strange looking "fox" well. (see photo).

"It was 1986 when this guy from Hughes Brook saw three foxes on the Camp 10 Road and he hit one with his truck. He brought it home to my son, a taxidermist, to have it mounted. At first they thought it was a fox, but then they discovered it was a coyote pup about eight weeks old. That fall, I went in on the north shore of Deer Lake on Camp 10 Road and set twelve fox snares. I went back three days later and I had caught three coyotes. These weren't pups, they were all yearlings and all males. There were lots of sign of coyotes," Jennings recalls. (see photos).

Jennings' catch marked the first time coyotes had been trapped on the island. The animals were eastern coyotes (Canis latrans) cousin to the western coyote.

According to Parker, the name Canis latrans means barking dog and was first given to the prairie coyote by Thomas Say in 1823.

Mike McGrath is a Fur-Bearer Biologist with the provincial government. He speculates on how the coyotes could have multiplied so quickly. "All it would take is one breeding female and they're pregnant in March (when the three animals were seen off Marches Point) and have their pups in May. If a pregnant female was one of the three on the ice off the Port au Port Peninsula we could have went

2

from 3 coyotes to 7 or 10 coyotes immediately. But that's assuming these three were the first ones. There could have been more here before them," McGrath says.

Jennings agrees with McGrath that coyotes could have been here long before the three travelers were seen off Marches Point. "In my opinion they were here a while before the first one was seen or caught. I had seen tracks I couldn't identify, but they weren't fox or lynx. If buddy didn't hit that one he took out to my son to have mounted, we might not have known for a few more years," Jennings says.

With the confirmed presence of coyotes in Newfoundland, a remarkable colonization had been completed. The coyotes were originally an animal of the southwestern American plains. Yet they had gradually expanded their range, invading the northeastern United States and Canada. Like a barbarian horde, the coyote was a relentless, unstoppable invader.

Shane Mahoney is a senior biologist with the provincial government. He describes the arrival of the coyotes as the most significant ecological event in Newfoundland since the introduction of the moose in 1904. "All islands have a different ecological system. Newfoundland has relatively few prey species and many predators. In every system coyotes are in, they have contributed to the mortality rates of a variety of prey species," Mahoney says.

... **CHAPTER TWO** ...

Were Coyotes Introduced?

Like so many wildlife issues, the establishment of coyote populations on the island of Newfoundland has generated controversy and dispute. But not only is there disagreement over how to manage or control the coyote, there is the assertion from some quarters that the provincial government had a hand in introducing the coyote to Newfoundland.

Eric Patey is a big-game outfitter on Newfoundland's rugged Northern Peninsula. He's been outfitting non-resident hunters for close on twenty years and he knows the country and its animals like the back of his hand. Patey has strong opinions about the coyote. "I think the government brought the coyote here. They say the coyote came across the Gulf, about 90 miles, yet it's only nine miles across the Strait of Belle Isle to Southern Labrador but no wolf has come across. I question it," Patey remarks. He thinks that government bureaucrats believed that Newfoundland lacked a predator on moose because the Newfoundland wolf was extinct. Similarly, the pulp and paper companies were complaining about moose eating all the trees. He thinks the combination of civil servants and large companies had a shared goal of controlling spiraling moose herds and so introduced the coyote to replace the

Newfoundland wolf which had become extinct sometime around 1922.

"I honestly believe the coyote was brought here. We went for one hundred years and then the coyote showed up. Wolves are only nine miles away in southern Labrador, and we get many white foxes from Labrador, how come another pair of wolves didn't come across," Patey asks. Patey is not alone in his assertion. Wayne Holloway runs a series of outfitting camps in the Middle Ridge area of south-central Newfoundland. He too believes the coyote had some help in coming to Newfoundland. When asked about the first coyotes that came to Newfoundland, Holloway replies "there's a lot of people, myself included, who would contest that claim. They may have been seen on the ice but that's no evidence they came from Nova Scotia," he says. Holloway thinks there's a human element to the coyote's presence here in Newfoundland that just hasn't been unearthed yet.

Similarly, sheep farmer Wade Alley of Robinsons on Newfoundland's west coast also thinks the coyote was introduced to the island. "I believe the government brought them in the early 1980s to kill off caribou sick with brain-worm."

Trapper and hunter Monroe Greening of Clarenville also believes that the government brought in the coyote. Greening says that he heard a story that coyotes were introduced to Newfoundland. "It was brought up at a trapper's convention in Montreal. A trapper from either New Brunswick or British Columbia asked another trapper how the coyotes were doing in Newfoundland. This guy said he helped catch the animals that were sent to Newfoundland. There's no proof or evidence to support this, it's just hearsay. But I believe the coyote was brought in like other species were," Greening says.

Greening's point is a good one. It is known historically that Newfoundland has experienced various wildlife introductions, some successful, some not. Sir Wilfred Grenfell's reindeer herd, the mighty moose, and snowshoe hare are three examples of introduced species. Wildlife Officer Bill Green also says that some Americans released mountain lions into the Long Range Mountains years ago and he still gets occasional reports of big cat sightings.

There is evidence that coyotes have been introduced into other jurisdictions. For example, biologists Gary Moore and Gerry Parker's research on these animals indicates that "there were a number of releases by sportsmen in states such as Florida (12 in 1925, 16 in 1930) and Georgia (6 in the late 1930s) .... Those and later releases have been identified as facilitating colonization of the southeastern states." Gerry Parker is retired from the Canadian Wildlife Service and is an authority on the coyote.

Given the history of animal introductions to Newfoundland, and given the fact that coyotes were inserted in other jurisdictions, it seems reasonable to ask if the coyote was intentionally introduced into Newfoundland.

This question was posed to a variety of provincial government Wildlife personnel. Bill Green, who responded to that report of wolves on the ice in March 1985, does not think that coyotes were brought to Newfoundland. "I heard the idea that coyotes were introduced by government but I know for a fact that coyotes were not brought here by the wildlife division," Green says.

Biologist Mike McGrath shares Green's opinion. "Well, if the coyote was introduced to Newfoundland, I guess that someone must have brought them to New Brunswick, Nova Scotia, Quebec, Ontario, New Hampshire, Vermont and all the other northeastern United States. If you track their range expansion it's very logical that they showed up here," McGrath says.

Another factor McGrath uses to debunk the idea that coyotes were introduced has to do with the fact that wolves were native to the island of Newfoundland. "I would say the extinct Newfoundland wolf would be a better scenario if someone was going to introduce something. You'd think they'd introduce wolves, not coyotes," McGrath says. The Biologist also does not think it is unreasonable to believe that coyotes crossed over from Nova Scotia, possibly Cape Breton, via sea-ice. "When you look at when they were reported, there was almost one hundred percent coverage (by ice), you could have walked across yourself. It was a heavy ice year and they could have walked across in a couple of nights," McGrath asserts.

The fact that three animals were seen off Marches Point in mid-March also coincides with the time of year when seals would have been in the Gulf, providing prey to draw the coyotes onto the ice. McGrath thinks that the open expanse of ice would not have deterred the coyotes, because they are essentially an animal of the plains, and unlike some creatures, are not afraid of wide open spaces. "So coyotes would be quite comfortable in that environment. It's certainly not beyond them being out on the ice," McGrath says. He also points out that the crew of the ferry boats that run from Newfoundland to North Sydney, Nova Scotia, reported seeing coyotes out on the ice in the 1990s. The Cabot Strait between Newfoundland and Nova Scotia is about 100 kilometers across.

Gerry Parker's 1995 examination of the eastern coyote suggests that coyotes probably crossed over from Cape Breton to Newfoundland. Parker's work in the study of the coyote is well-regarded.

Wayne Barney is Species Management Co-ordinator with Newfoundland's Inland Fish and Wildlife Division. He bluntly states that the idea that coyotes were introduced is absolutely false. "Since 1985 we've had subsequent reports of coyotes on the sea-ice (between here and Nova Scotia). For example, we had reports in 1999 and again in 2000," Barney says.

Wayne Barney grew up in the Labrador Straits and he knows that area well. He thinks he knows why wolves haven't crossed over from southern Labrador into Newfoundland. He says that even though southern Labrador is only nine miles from the island, the sea-ice is much more inhospitable than the ice in the Gulf of St. Lawrence. "The ice in the Straits moves quickly with the tides. Besides, wolves are not known to be pack-ice animals. And, in southern Labrador, wolf densities are not that high," he says. Barney also points out that Newfoundland is not the first island coyotes have conquered. The cunning dogs also invaded Cape Breton Island and Prince Edward Island.

Randy Dibblee is a wildlife biologist with the Government of Prince Edward Island. He has received reports of coyotes crossing the Northumberland Strait between New Brunswick and PEI on sea ice. (see photos).

John Blake also dismisses the notion that coyotes were introduced. "Secrets are hard to keep, especially in government." Blake says that if there was a government sponsored conspiracy to introduce coyotes, it would have leaked out by now. While no hard evidence can be found to support the line of thinking that government introduced coyotes, the fact that some individuals believe that government would secretly bring in these predators highlights a general lack of trust in the machinery of government by some members of the public. Wayne Holloway and Eric Patey are outfitters or small businessmen who employ many people in their camps. Monroe Greening is a retired railroad engineer and Wade Alley is a hard-working farmer. These men are all responsible citizens and the fact that they believe some secret project was hatched to introduce the coyote underscores a problem facing wildlife managers. Some members of the public appear to lack trust in the government.

One thing is certain though, the arrival of the coyote in Newfoundland in the mid-1980s has to be placed in the context of the animal's spread across eastern North America throughout the twentieth century.

... **CHAPTER THREE** ...

History of Colonization of Eastern North America

A variety of researchers have documented the historic range of the coyote and its gradual spread eastward. For example, Robert Wayne and Niles Lehman say that, prior to the arrival of the Europeans in North America, the coyotes' range was restricted to the great plains of southwestern North America, stretching from present day Canada to Mexico.

Perhaps the best known coyote researcher is retired Canadian Wildlife Service Biologist Gerry Parker. Parker and co-author Gary Moore trace out the coyote's eastward trek in the nineteenth and twentieth centuries.

Historically, the coyote occupied the prairie regions west of the Mississippi River. Throughout the late 1800s much of the western prairies was turned into pasture for livestock as the human population spread westward. This was accompanied by a decline in large predators such as the grizzly bear, wolf and mountain lion. As its larger competitors were removed, the highly adaptable coyote expanded its range and increased its numbers in the face of the European occupation of western North America.

In the late 1800s, the coyote followed the gold rushes into the Yukon and Alaska, drawn by clearings, waste and garbage left by prospectors.

The period from 1900-1939 represents the first coyote colonization to the eastward. Coyotes followed the south shore of the Great Lakes into upper Michigan and Ontario. "Coyotes reached southwestern Ontario by 1919, southeastern Ontario by 1928, New York by 1936 and Maine by 1938."

From 1940 to 1959 the coyote continued expanding throughout the northeastern United States and southern Quebec. In the late 1950's the first coyotes began appearing in Massachusetts. However, population densities remained small as the coyote continued to disperse. Through the 1960's, 70's and '80's the coyote continued its eastward march like a conquering army. Where the mighty wolf had been brought to the edge of extinction by loss of habitat, poison, traps and guns, the wily coyote flourished in the face of a growing human population and massive control programs. For example, in the state of Mississippi, 500 coyotes were harvested in 1975 and the harvest grew to 40,000 by 1988. Estimates now place Mississippi's coyote population at a staggering 400,000 animals, although it is important to remember that increased harvests are tied not only to increased densities but also to increased hunting and trapping efforts.

Similarly, in Maine, about 300 coyotes were taken in 1977, and by 1987 over 1,600 were harvested. Parker (1995, p.80) states that research seems to indicate that coyote densities are higher in the southern U.S., than in the north. Parker states: "There is also reason to believe that densities of coyotes may increase from north to south, probably due to an increase in food availability and a decrease in environmental stress."

The first coyotes began showing up in New Brunswick in the late 1950's. According to Parker, the first confirmed coyote in Atlantic Canada was shot by Bob McFarlene at Sussex, New Brunswick on Boxing Day 1958. But up to 1972 only three more coyotes were known to have been killed in New Brunswick. Parker and Moore assert that these animals probably came from Maine and the Gaspe region of Quebec.

Wherever it originated, the eastern coyote soon spread like a wind-whipped wildfire across grassland. In 1974, 5 coyotes were trapped in New Brunswick. In 1977, 30 pelts were exported for sale and by 1988 1,633 coyote pelts were shipped to market.

Cade Libby is the Fur-Bearer Management Biologist for New Brunswick and he says that while no formal population estimate has been made, "we expect our population of coyotes to be similar to Maine's. They have 10-15,000 coyotes and we can expect that here. The population seems to be stable since the mid-1990s."

Around the same time that the coyote was increasing its densities in New Brunswick, it was also invading Nova Scotia. The first coyote specimen collected in Nova Scotia was trapped in 1977 by Howard Porter. It was the sole coyote harvested that year. By 1989, 581 coyotes were taken. Cape Breton Island gave up its first coyote pelt in 1981-82 and it is speculated that it probably crossed over on ice in the late 1970s. Today, the coyote is well established throughout all of Nova Scotia. According to the director of wildlife for Nova Scotia, Barry Sabean, there are now estimated to be 8-10,000 coyotes in the province today.

In 1983, a male coyote was trapped on the eastern portion of Prince Edward Island by Johnny MacCormack. It is thought the animal crossed the ice from northern Nova Scotia, a distance of 12 km. By 1991, 17 coyotes were harvested on Prince Edward Island. In 1994, Parker states that 131 coyotes were taken on little PEI, "indicating a rapidly expanding population in that province." Randy Dibblee is a wildlife biologist with the government of Prince Edward Island. He says that from 2002-2004, a total of 1,375 coyotes were taken on the island, approximately 450 a year. "So we're probably trapping 25-30 percent of the population, so we probably have 1,500 to 2,000 animals here. But that's just a ballpark figure," Dibblee says. He adds that he thinks that PEI's coyote population has now "maxed out."

An interesting tidbit of information provided by Dibblee is that he has reliable reports that coyotes colonized the

Magdalen Islands in the Gulf of St. Lawrence about two years ago. The Magdalen Islands are the latest in a string of islands colonized by coyotes. Others include Cape Breton Island, Prince Edward Island and, of course, Newfoundland. John Blake points out that the colonization of this string of Atlantic islands should defuse the conspiracy theory that coyotes were intentionally introduced into Newfoundland.

With regard to this province, we know that three wolf-like dogs were seen on the ice near Marches Point on the Port au Port Peninsula in 1985. A coyote pup was killed by a vehicle on Camp 10 Road near Deer Lake later and trapper George Jennings caught three coyotes shortly after that road-kill.

Parker and Moore state that the most plausible source of coyotes in Newfoundland is northern Cape Breton in winter-time. Parker also suggests this in his 1995 book Eastern Coyote.

Biologist Mike McGrath says that the government has no idea of the numbers of coyotes in Newfoundland today. "All I can say is that their numbers are still increasing dramatically and we still don't know at what level they'll level out at," McGrath says. He estimates that in 2003 the number of coyotes taken was between 100 and 300.

McGrath explains that getting an abundance figure is not really relevant since coyote populations haven't yet stabilized. "So putting a concerted effort into density estimates is not something that we're putting a large amount of emphasis on," he says. Big game outfitter Wayne Holloway wonders why more effort hasn't been put into estimating the size of the coyote population; estimates are done of other species, such as moose and caribou, and recognized models exist for doing population estimates. Holloway believes that coyotes now number in the range of 40 to 50,000 animals. He says that he has come up with this figure using a conservative model to estimate the growth of coyotes on the island.

He factored in things such as the assumption that one of the first animals seen on the ice was a pregnant female, the fact that coyote breed at age one, that the average litter size is four to six pups and that they may produce more pups than this in accordance with the health of the environment. "Where they have a super food supply, like here in Newfoundland, litter sizes could be larger," he says.

Mike McGrath shakes his head when asked if coyotes could number 50,000 animals. "No, that's totally, totally erroneous. If there were 40 to 50,000 coyotes on the island we wouldn't be harvesting 7,000 foxes because coyotes would have replaced foxes. If there were 40 to 50,000 coyotes, that would put the density at around one coyote per 2 square kilometers. Therefore, they haven't reached those kinds of densities. Whether or not they will, I don't know," he says.

Similarly, Shane Mahoney says that for there to be 40 to 50,000 coyotes on the island, a very rapid population growth would have had to occur. Although he admits rapid population expansion is possible with coyotes. "We do not know things that would help us establish the size of the coyote population, such as home range size and litter size. They can throw out big litters but we don't know if they're doing that here. We've never radio collared a coyote and we don't have any basic biological information on the animal."

The Newfoundland case aside, Parker and Moore say that "today the coyote is found in nearly all of the continental Unites States and all Canadian provinces and territories." They theorize that the coyote's domination over North America cannot be separated from the human colonization of the same territory. As humans pushed westward they converted forest to farms and opened prairies to pastures. Railway lines, roads and mines were opened. Big carnivores like grizzlies and wolves decreased. The adaptable little dog began to spread. Co-existing with humans and often living in close proximity to people, the coyote marched eastward as ranchers, farmers, lumberjacks and miners tromped westward. Millions of coyotes have been shot, trapped, poisoned and run down. Yet the resilient little wolf marches on like an unstoppable army.

Much has been written and said about how cowboys won the American west. A tale as great is how the coyote won the north and east. The biological make-up of the barking dog has aided its range expansion.

...CHAPTER FOUR...

Coyote Biology

Eastern coyotes have good vision, hearing and sense of smell. They are remarkably adaptable and are true omnivores. That is, they can and will eat everything from big game to human garbage to berries. Free-lance writer Leslie McNab states its "reputation for adaptability fits with the coyote's mythic status as a trickster." North American Indians viewed the coyote as a playful god.

The Eastern coyote's origin may have involved breeding with gray wolves. Researchers Robert Wayne and Niles Lehman did mitochondrial DNA analysis on 350 gray wolves and 327 coyotes. They concluded that the two species had inter-bred during the coyote's range expansion eastward where coyotes and gray wolves shared the same range at low densities. It is stated that hybrids (i.e. wolf-coyote cross) are fertile and able to breed.

Typically, in the dog world, wolves will dominate coyotes and coyotes will in turn dominate foxes. Wolves have been known to chase and kill coyotes on sight, and coyotes have been known to avoid the range of wolves. But in some cases, coyotes may actually follow wolves to prey on remains left in the path of the powerful predator. For example, Parker points out that coyotes scavenge on moose carcasses left by wolves.

It seems that in cases where wolf and coyote densities are low, and potential mates few and far between, the two species have inter-bred. Parker says that, as a consequence, the eastern coyote has developed a larger body size and a social nature more common to wolves. These traits would help the coyote in pursuing larger prey, such as white-tailed deer or caribou.

An important part of the coyote's link in the ecological system of northeastern North America is that, by the early twentieth century the wolf had disappeared from much of the forest, leaving a void that coyotes would eventually fill.

According to the research of John Litvaitas of the University of New Hampshire, coyotes and members of the cat family such as the lynx will co-exist though separated "by differing morphological adaptations to snow." For example, Litvaitas says that felids like lynx "often rely on dense understory cover to facilitate their stalk and ambush style of hunting, in contrast to the open-area pursuit methods of canids." That is, lynx will hunt snowshoe hares in thick low growth, while coyotes prowl more open country.

According to the National Trapper's Association (NTA) Handbook, the eastern coyote attains heavier body weights than its cousin, the western coyote. Some males have attained weights of 60 pounds, but the majority weigh around 30-35 pounds. Parker suggests similar weights in his book about coyotes. "The occasional old male eastern coyote might reach those weights (60 or 70 pounds) as well, but this would be uncommon." Pictures of coyotes in the 60 pound weight range bear a striking resemblance to wolves.

New Brunswick Fur-Bearer Biologist Cade Libby says that coyotes harvested in that province have varying weights, some weighing as much as 53-54 pounds. According to Parker, sexual dimorphism or differences between the sexes is the rule among canids. "Male coyotes are larger and heavier than female coyotes throughout its range." John Blake points out that this is also true for other canids such as the house dog or fox. Parker states that the average adult male coyote weighs about 35 to 40 pounds, while adult females generally weigh 5-10 pounds less.

In Newfoundland, Mike McGrath says that the biggest coyote he has studied weighed in at 43 pounds. "They stand about 24 inches at the shoulder and from a distance look like a German Shepherd dog," he says. McGrath goes on to say that some people have reported seeing 80-100 pound coyotes and he thinks that this is related to the fact that coyotes look bigger than they actually are. John Blake says that he thought the first coyote he shot in the winter of 2002 would tip the scales at over 50 pounds. However, the animal weighed only 38 pounds. Kim Bridger is a biology graduate student who is studying coyotes' diet in Newfoundland. She, too, says, "when you see them in the field they look larger, with their coat, than they actually are." Sheep farmer Wade Alley killed a large male coyote near his barn in Robinson's that weighed close to 70 pounds. Alley, who is well-used to being around animals, is confident of his estimate. Retired wildlife officer and environmental consultant Bruce Porter has seen many coyotes from helicopters as he flies over the barrens around Granite Lake on Newfoundland's south coast. "They vary in size. Some are quite large, as big as a German Shepherd dog, while others are smaller." Similarly, big game guide Chad Snow of Lewisporte has seen a few coyotes in his travels and they seem to be "about the size of a German Shepherd."

Parker says that coyotes collected from Vermont and Massachusetts in the mid-1970's had body lengths that varied from 43 to 59 inches (including tail). A study from Arkansas found that coyote tails measured from 11 - 15 inches. Coyotes have 42 teeth including four long incisors. Their eyes are yellow or amber with circular black pupils. Coyotes are, in the words of trapper Eugene Tiller, made for traveling. In fact, the dogs are capable of speeds of 35 mph for short periods of time. Most coyotes are a mottled grey color with a lighter belly, although brownish or reddish colors are also common. Black is a less common color, says the NTA Handbook.

The guard hairs on a coyote's back are about three inches long, while hairs in a patch between the shoulders, known as the mane or hackles, are about five inches long. The animal's thick, bushy coat contributes to the visual effect of making it seem bigger than it really is.

There is some evidence that suggests coyotes are monogamous and mate for life. But Parker says that "as a rule, coyotes do not mate for life. Some pairs may remain together for a number of years, however."

Robert Chambers' research shows that mating occurs from late January through February, although the NTA Handbook suggests that in northern climates breeding may occur in March. Gestation lasts around 63 days and most litters are born in mid-April. The females usually give birth in an underground den, and often the same den is used year after year. They may also use caves, hollow logs, or even culverts as dens. The young pups don't open their eyes for about their first 12 days and the mother stays with them during this period. The male brings food to his mate while she is nursing the pups. The nursing period is about six weeks long.

Litter sizes average 5-7 pups in many areas. The NTA Handbook says that litters may average 8-9 pups where there are few coyotes. However, this may reflect the fact that there is more food and the coyotes are healthier. Parker says that "in times of plenty, litters of 9-12 are not uncommon." He goes on to suggest that some females are known to have huge litters from 17-19 pups, but this is the exception rather than the rule. Parker points out that "mortality of pups can be high" especially if the female is nutritionally or socially stressed.

Female coyotes will breed at eight or nine months of age, but many pair up with a mate when they are 20-22 months old. Parker says that generally "60-80 percent of adult females normally breed and bear young each year. "Robert Chambers' research shows that there is no decrease in fertility as a coyote ages and "productivity of eastern coyotes 5 - 8 years old was as great as that of younger adults." Some research has suggested that litter size may actually increase with age.

Chambers concludes that eastern coyote reproduction is responsive to "changes in population and environmental conditions." As noted previously, litter size may actually increase as the population density of coyotes decreases. This may be related to more available food, thus allowing breeding pairs to be healthier. Shane Mahoney explains that coyotes can, under some circumstances, increase litter sizes in the face of increasing hunting pressure. How this occurs is that with a fairly dense coyote population there is a competition for food. When the population is reduced there is more food available which means healthier coyotes which in turn equals bigger litters. Like so much about the Newfoundland coyote, Mahoney says that scientists are unsure of litter sizes on the island.

John Blake says that coyotes have an intricate social structure and dominant males and pairs will have an established home range that, during certain times of the year, will be defended quite rigorously. If such pairs are removed through hunting, this allows neighboring coyotes to produce more young and fill the void. Therefore, territoriality, along with food supply, is critical in establishing population density models.

Robert Chambers writes that "attempts to achieve long-term reduction of eastern coyote numbers by liberalizing harvest would likely fail as a consequence of the coyote's ability to make compensatory reproductive responses." That is, attempts by humans to eradicate coyote population may actually lead to females having larger litters.

Coyotes generally do not have a long life expectancy, however, some research indicates that they can be long-lived. They are very rarely killed by other wild animals, apart from wolves; the coyote's chief predator are humans. When not killed by man, coyotes can reach ages of 14 or 15 years old.

Coyotes are incredibly social animals and communicate vocally with barks, howls and yips. To hear the wolf-like howl of a coyote just before dark in a remote area is a spine-tingling experience not easily forgotten.

According to the NTA Handbook, the territory of coyotes varies in size depending on food supply. Generally, it appears that females have smaller ranges than males. Females may range over ground of 5-8 miles in radius, while males may have a territory as large as 30 miles that overlaps the range of other coyotes. Parker says that coyote's home ranges are "usually largest in the northern spruce-fir forests in winter." It seems that the coyotes in northern forests do more roaming to find prey during harsh northern winters.

John Blake uses the example of the Newfoundland pine marten to expand on the discussion of home range size. Blake says that martens in Newfoundland have home ranges three times the size of those in Labrador or Mainland Canada, likely a result of the small prey base available. He says that it is quite possible that coyotes in insular Newfoundland will have very large home ranges, possibly several times larger than coyotes in other jurisdictions. Blake suspects that in most of Newfoundland the coyote's huge home range size will be a limiting factor in its ability to establish the extremely high densities common in places like Western Canada.

According to research from central Alberta, "the home range of an adult coyote averages 12 square kilometers, but can be two or three times larger." This same research by Acorn and Dorrance says that "adult coyotes tend to establish home ranges while juvenile coyotes tend to wander before selecting a permanent territory." Parker goes on to explain that while coyote pups usually disperse from their parents' range in fall or winter, pups may delay dispersing in the northeast "where winter survival can be enhanced through cooperative hunting of larger prey by the family unit." This may be the case in Newfoundland, where there are reliable reports of packs of coyotes hunting caribou and moose in the snow.

Radio-collar studies of coyotes in other jurisdictions have found that they are most active and travel the greatest distances between 6 p.m. and 6 a.m., although travel times will vary seasonally. For example, during the coldest part of winter, small mammals like hares will wait until the heat of the sun makes foraging more energy efficient. During such times, coyotes will time their movement with those of their prey thus becoming more active during late morning. This is also true of fox.

This study found that the average daily distance traveled was 5 km in fall and 8 km in winter. One animal journeyed a total of 140 km between November and January. John Blake says that he has tracked coyotes that have traveled as far as 15 kilometers in one evening. No coyotes have yet been radio collard in Newfoundland and so there is no data on their home range size. However, trapper George Jennings knows first hand how far coyotes can travel and how powerful they are for their size. Jennings has been trapping for 40 years and he lives in Hughes Brook in the Bay of Islands. He had a coyote break off a fox snare he had set on Deer Lake's north shore and that animal was later killed by a car near St. Anthony with the snare still around its neck. "I knew my snare from the way I had it tied," Jennings said.

Similarly, big game outfitter Eric Patey has been prowling the hills of the Northern Peninsula all his life. He has killed several coyotes and has witnessed them killing caribou. He knows the power, speed and agility of these animals. "I've seen them come out of nowhere and go running after caribou. They can jump about twelve feet in full stride," Patey says.

Coyotes do have a social order and certain animals will be dominant over the rest. Also coyotes, like wolves, are known to form packs. Daniel Harrison, of the University of Maine, has conducted research on coyote social relationships which shows that coyote families maintained ranges exclusive of other family groups. This research suggests that pack formation amongst coyotes may be related to large size of prey (i.e. big game) and/or clumped distribution of food which would promote delayed dispersal of the pups from their mother. Pack formation may help the coyote in preying on large game and/or defending such prey. Anecdotal evidence from reliable sources indicates that coyotes are hunting in packs in Newfoundland.

For example, Brent Sellars is an ecologist with Newfoundland and Labrador Hydro Corporation. Part of Sellar's job involved monitoring the Granite Canal hydro project on Newfoundland's barren south coast. Sellars regularly uses a helicopter to fly over the area and he has witnessed packs of coyotes hunting. "We see more groups of two or three than singles. We see them in the winter when there are lots of caribou in the vicinity of the project. We tend to see them on the edge of the lake or on open bogs," he says.

Sellars tells of one incident he and his helicopter pilot witnessed on the northeast corner of Granite Lake. The helicopter was just crossing over the lake and coming onto a bog on the pond's edge when they saw a lone coyote pursuing a caribou eastward across the open marsh. "He was biting at the caribou's hind leg and it was bleeding. Two or three times the caribou turned around and faced the coyote and drove it away. But whenever it turned back, the coyote would close in again and start tearing at the hind quarter. At first we thought there was only one coyote, but then we saw that there were two more hanging out on the edge of the bog. It seemed like the one chasing the caribou was trying to drive the animal towards the other two coyotes. But eventually the noise from the chopper drove the coyotes away," he says.

Bruce Porter also says that he has witnessed groups of coyotes working in relays to bring down adult caribou.

Another example of a coyote pack was related to me by a life-long friend and avid hunter, Ron, who did not want his full name used. Ron and a buddy were moose hunting in Sheffield Lake a couple of years back. It was late in the fall with snow on the ground. They came upon a moose just before dark. They had a shot but the moose bolted.

Blood on the snow told the men their prey was wounded and with darkness falling they returned to their camp to get flashlights.

Armed with flashlights, they followed the blood trail. The track took them down through a valley, across a big bog and toward a stand of trees. As they neared the trees they noticed what seemed to be several sets of dog tracks in the snow. As they got closer to the trees they heard growling and snapping. Shining their flashlights around they saw four or five sets of eyes reflected in the beams. Since it was dark, and they were armed with only one rifle between them, the men retreated and decided to return in the morning.

They came back early the next morning, armed with a shotgun and a rifle, and tracked their way to the moose. Or what was left of the moose. The moose had been torn to bits. The culprits had moved off, but the tracks in the snow confirmed that several coyotes had devoured the mortally wounded moose.

A similar story was related to me by big game guide Chad Snow. Snow says that an out-of-province sportsman he was guiding knocked down a caribou stag just before dark on a Saturday evening. They cleaned up the stag and, because it was getting dark, decided to leave it until the next morning to quarter and bring back to camp. When the guide and hunter returned in the morning, there were four coyotes feeding on the caribou. "And they didn't mind us too much. They just hung around there." Needless to say, the meat was ruined for human consumption.

Besides forming packs to hunt big game, coyotes have several other traits that help them flourish. Laitivis's study suggests that the coyote will eat just about anything, have large litters in a relatively short time, is able to adapt to a variety of habitats, and can co-exist with humans. For example, according to the NTA Handbook, thousands of coyotes live within the limits of Los Angeles. Likewise, Shane Mahoney has seen coyotes in Vancouver.

As mentioned above, coyotes are omnivores or generalists. That is, they can and will eat anything, from big game to vegetation. But a large part of their diet is prey that the coyote "stalks, chases, captures and kills."

The research of Peter Pekins of the University of New Hampshire shows that in "northern, contiguous forest habitats" white-tailed deer and snowshoe hare are the winter dietary staples of coyotes.

In regions where there is more agriculture, the eastern coyotes' diets are more diverse. According to Guy Connolly of the U.S. Department of Agriculture, "coyote predation is a well-recognized cause of livestock loss in North America." Most commonly, coyotes prey on sheep, lambs, calves and goats. More will be said about the relationship Newfoundland's sheep farmers have with coyotes in a later chapter.

Not only will coyotes prey on livestock, but they have even been recorded as eating watermelons on farms in Alabama, Texas and Oklahoma.

According to Prince Edward Island Biologist Randy Dibblee, there are no white-tailed deer on the island and coyotes eat whatever they can, including skunks and racoons.

Kim Bridger is a graduate student who has been studying the diets of Newfoundland's eastern coyotes by dissecting and analyzing carcasses. "The majority of dietary content from coyote stomachs is mostly caribou and snowshoe hare. There's a little bit of supplementation of diet with birds of all species. I have also found various berries in their stomachs. They eat a lot of vegetation because it acts as a laxative and helps them digest the high amounts of protein (i.e. meat) they're taking in."

Bridger's research is important because it demonstrates the coyote's apparent reliance on caribou, which was not predicted by biologists such as Parker. In fact, Parker had theorized that "snowshoe hare will be the main source of food for coyotes in Newfoundland, especially in winter." But Parker did point out that coyotes may develop modifications in "body size and behavior to take advantage of food sources available."

One very interesting point about the relationship between predator and prey raised by Parker is that larger prey may equal larger predators. That is, coyotes may evolve physically into a larger size because of their use of larger prey such as moose and caribou.

John Blake says that black bear predation on caribou is well-documented in Newfoundland. Such predation could account for the huge sizes black bears are known to attain in Newfoundland.

To conclude on diet, Parker says that "one reason for the rapid colonization by the coyote of most of eastern North America is its ability to adapt its feeding habits to the available food supply."

While there has been some fear associated with the arrival and dispersal of coyotes in Newfoundland, coyote attacks on humans are rare. Parker says that the first and perhaps only known fatality was a three-year-old child killed by coyotes in Los Angeles County, California. This attack occurred in 1980. The coyotes involved had become used to food handouts and were habituated to living near human dwellings situated near mountain canyons.

Parker discusses four cases of coyote attacks on children which produced major injuries and ten attacks which resulted in minor injuries. The four cases of major injuries involved children under the age of five years in which coyotes attacked the head/face/neck area of the children much like coyotes go for the head/neck of sheep.

For example, on August 30, 1985, a 4 year-old girl was playing in a campground in Jasper National Park. She was out of sight of her parents when she was attacked by a coyote. The child's screams attracted the parents who drove off the coyote. The child received numerous wounds to the face.

Earlier that year a coyote attack on a child had occurred in Jasper townsite. A mother went to check on her two-year old daughter, who was playing in the garden, to discover a coyote dragging away the child's limp body by the throat. The parents drove off the coyote and the child was found to have extensive wounds to the face, neck and throat, although she did recover.

Parker relates another story of serious injuries from a coyote attack in Creston, British Columbia. Two sisters aged 10 and 18 months were walking along a path near a highway stop. A coyote attacked the smaller child, knocking her down and biting her head and neck. The father rescued the girl and the police later shot the animal which remained in the area. The child needed over two hundred stitches. Undoubtedly, coyotes have proven they can and will attack children and are quite capable of maiming and killing them.

Parker also says that there have been several cases in Cape Breton Highlands National Park of people being bitten or "nipped" by coyotes. Parker concludes:

> *the presence of the eastern coyote throughout the northeast does present a potential risk to humans, especially children, but that risk is slight when weighed against other dangers to human health...intentional encounters, such as feeding or close photography should be avoided. Small children should not be left unattended when in wilderness areas, especially in campgrounds where coyotes may be especially bold...The public should not consider the coyote reason for avoiding wilderness areas, but rather should treat the coyote with caution and respect due all wild animals, especially those that have the potential to cause bodily harm and personal injury.*

Connolly's research cautions that coyote/human interactions can be expected to increase in eastern North America as both the coyote and human populations grow. (More about this in a later chapter).

Information obtained from an internet search using the phrase "coyote attacks on humans" revealed many web-pages on this topic. One such web-page states that the California Department of Fish and Game estimates that one person per year gets bitten by a coyote in California. In the state of Arizona, there were reportedly seven coyote attacks between 1993-1997, with over half of them occurring in 1997.

However, to put coyote attacks on humans in perspective it is worth noting that the United States Humane Society 1999 report says that 300 people were killed by domestic dogs between 1979 and the late 1990's.

Sheep farmer Howard Morry says that he met a female tourist from British Columbia who showed him the scars on her lower leg from a coyote bite. Morry says that the woman told him how the coyote in question was trying to drag away one of her grand-children and when she attacked the animal with a stick it bit her. Fortunately, the coyote did retreat.

Mike McGrath is the Fur-Bearer Biologist with the Government of Newfoundland and Labrador and he points out that, given the millions of coyotes that inhabit North America, coyote attacks on humans are very, very rare. McGrath says he is familiar with the case of the child killed by coyotes a few years ago in California. In that situation, the child was mauled by coyotes that had become accustomed to humans, who were actually feeding them on a regular basis.

McGrath says that like bears in National Parks that get fed by visitors, the most dangerous wild animals are those that become too familiar with humans. "If you look at it in perspective your risk of getting bitten by a dog-like animal is much greater if you go for a walk to the post-office versus going in the woods," McGrath says.

Similarly, John Blake, Manager of Conservation Services with the Inland Fish and Wildlife Division, says that there is a "very, very, very small potential for harm to humans." Blake says that the data to support the argument that coyotes will regularly attack humans is just not there. "There has been an incident or two over the years across North America, but moose are much more prone to cause bodily injury than other animals," Blake says.

Shane Mahoney concurs with McGrath and Blake. "I think the risk is extremely low. But in circumstances where the animal is cornered or habituated to people a small child could be vulnerable. People should treat the animal with respect."

Biologist Randy Dibblee says that there have been no coyote-human interactions since the coyote colonized PEI in 1983.

Unfortunately, coyotes have a long, bloody history of interacting with sheep.

... **CHAPTER FIVE** ...

Coyote and Sheep

A s Connolly's research shows, coyote predation is a well-recognized cause of livestock loss in North America. For example, it is estimated that coyotes in the United States killed sheep, lambs and goats valued at 16 million dollars in 1990. He writes that "the predominant coyote damage problem in the United States is predation upon domestic livestock, particularly, sheep and lambs." Parker says that the coyote's colonization of eastern North America would not have generated as much controversy except for the fact that coyotes do prey on sheep and everything from pigs to horses to fruit, crops and stored livestock feed.

While the coyote has only been in Newfoundland since around the mid-1980s, it is causing problems for the province's sheep industry. While some nature lovers may be thrilled to see a coyote, or to hear it howling, the farmer with sheep and lambs in the pasture reacts with anger, hatred and resentment.

Deborah Guillemette is the Executive Director of the Newfoundland and Labrador Federation of Agriculture and she is very familiar with the problem of coyote predation on sheep. "I first heard about coyotes five or six years ago. Now I hear about coyotes every day from our members. The fact I hear so much about it underscores the problem," Guillemette says.

Paul Dunphy is a livestock specialist with the provincial government and he, too, is aware of the coyote problem. Dunphy says that predation by coyotes on sheep became a problem around the early 1990's, then became a major problem in the late 1990's.

Pat Hewiit is a businessman and the Mayor of St. Shott's on the southern Avalon Peninsula. "Coyotes have been a big problem for the sheep farmers here since 1999. Some farmers are so frustrated they're considering quitting," he says.

Howard Morry is the President of the Sheep Producers Association of Newfoundland and Labrador (SPANL). He says that about three years ago close to 100 young lambs were killed in the St. Shott's/Peter's River area. Research from Alberta indicates that coyotes kill lambs in preference to sheep and "lambs comprise about 70 percent of the sheep killed by coyotes in Alberta."

It is worth pointing out that St. Shott's is about as far away as you can get from Marches Point on the Port au Port Peninsula. The two areas are at opposite ends of the island. The fact that coyotes have arrived in St. Shott's demonstrates that they have completed their expansion across the breadth of Newfoundland. In fact, the first place biologist Mike McGrath saw a coyote was on the pastureland near St. Shott's. "There had been a pack of three coyotes there and I shot video footage just after two had been shot. It was an adult female," McGrath says.

Cyril Myrick of St. John's told me that a coyote was witnessed near Cape Race lighthouse on the southern Avalon Peninsula, heading toward a flock of sheep that were about a mile and a-half away. This sighting occurred around mid-day.

The sheep industry generated three-quarters of a million dollars in sales in 2002 and Newfoundlanders consumed approximately 952,000 pounds of lamb. However, of that, only about 260,000 pounds was local lamb. Guillemette sees lamb consumption patterns as an opportunity for the industry to grow, however, she thinks coyote predation is holding the industry back. "Sheep are a great species for Newfoundland. They can be grown on marginal land and there are pockets of Newfoundland where sheep farming could expand. Also, sheep have lower start-up costs than dairy farming, which is very hard to break into. Sheep also have very low feed costs, they will eat grass on marginal land. The coyote is the major impact holding the industry back," Guillemette says.

Paul Dunphy agrees: "Coyotes are having an adverse effect on the sheep industry in the province. We produce only about 25 percent of the lamb we consume here. Coyotes are one problem that sheep farmers have to deal with to stay in business."

Howard Morry also thinks that there is a large potential for growth in the sheep industry. Morry sold about 500 lambs last year. However, he knows too well that coyotes are keeping the industry down. "Two years ago on the Gaskier's/St. Mary's pasture there were 80-odd sheep killed. That pasture used to have as high as 1200 sheep, now it has zero."

Morry says that the coyote is really affecting the sheep industry and many people are getting out of the business and are selling off their flocks. Morry recognizes that not all coyotes may be sheep killers, a point that Parker makes in his book. Morry says: "If you've got coyotes in your area and they're not killing your sheep, leave them alone. Because you might kill them and then coyotes that are sheep killers might move in."

In the early twentieth century, there were approximately 100,000 sheep kept by Newfoundlanders. That has declined to about 5,800 ewes and 7,300 lambs today. Guillemette says that Newfoundland could benefit from sheep production if something was done about coyote predation. "Most sheep farmers are in rural areas so it's an important economic activity. Whether the new shooting season will have an impact on coyote's remains to be seen," she says. Guillemette points out that sheep farming is extra income and could be used to help bolster the economy of rural communities.

Howard Morry knows some of the steps that local sheep farmers have taken to protect their flocks against coyote attacks. Some farmers use specially bred guard dogs, but Morry points out there are drawbacks with dogs. "I tried a guard dog, but I didn't have him castrated and he wanted to roam. And the dog has to be fed, so that's an added expense. Also, take the St. Shott's pasture, it's ten miles long. How many guard dogs would you need to protect it?"

Some of the various breeds that are used to guard sheep include the Akbash, the Anatolian Shepherd, the Great Pyrenees, the Komondor and the Maremma (see below). Some others, such as Morry and Sandy and Laurie Ruby, have brought in the South American llama which is "supposed to be hell on dogs and coyotes." The sheep will actually lie down around the llama at night (see photo). Morry keeps many of his sheep on two small islands near the community of Ferryland, thus avoiding coyote predation.

Paul Dunphy says that Joachim Ryan, in O'Regan's, uses a donkey to guard his flock, while Dick Whittaker has experimented with electric fencing. In some jurisdictions propane exploders are used to scare off coyotes, but Dunphy says that frightening devices are not used in this province. In Newfoundland, wildlife managers have adopted a strategy that they believe will help protect the province's sheep farmers.

Wayne Barney is the Species Management Co-Ordinator with the Inland Fish and Wildlife Division. He says that there is a hunting permit system in place to allow dealing with coyotes on a problem basis. For example, livestock producers can get a permit to address specific marauding, problem coyotes. "We can control the coyote population in specific areas like around a sheep farm. But we can't control the entire coyote population," Barney says.

Paul Dunphy says that the provincial government has done extensive training with sheep producers about preventing coyote predation. The department even brought in Alberta-coyote specialist John Bourne in 2001 to tour pastures in Newfoundland and speak to farmers about preventing loss to coyotes.

Wade Alley, like his father, has been sheep farming all his adult life in Robinson's on the province's west coast. Alley also has some cattle and pigs and grows vegetables. He has lost many sheep to coyotes. The coyotes are brazen and have come right onto his property.

Howard Morry says that the coyotes usually kill at dark or early morning. "They rip the throat out of the sheep and then open them up and go for the heart, lungs and liver."

Alley says that in 2002 he lost about 32 sheep to coyotes. "Thirty-two sheep at about $150.00 each. Plus, some of the sheep were heavy with lamb at the time, so it was quite expensive, I had very heavy losses," he says. Alley shot and snared four coyotes, two males and two females, that year.

Unfortunately, quantifying the total number of sheep lost to coyotes and their dollar value is difficult according to Paul Dunphy, there are no official figures on this.

The first trouble Alley had with coyotes was about six to seven years ago when he lost several newborn lambs in the spring. "They probably killed two or three lambs but I got them right away. I killed a female and her brood of four or five pups.

Today, Alley has about 75 sheep and he says this is less than he normally would have and it's all due to the coyote. He has also had to invest money in protecting his farm from further attacks. "I've got three guard dogs; Maremma's. They're big white dogs specially bred to protect sheep. They cost around $500.00 each. The coyotes are all around but the dogs keep them off. I have to do whatever it takes to be rid of them," Alley says. (See photo).

Alley is very frustrated with what he sees as a lack of government support for sheep farmers. "Wildlife will give you no help. If I can take care of it or not, they give no help. If you make it on your own that's fine and if you don't that's it. They've got a poor attitude," he says. Howard Morry agrees with Alley that the government is doing little to deal with the coyote problem or provide help to sheep farmers. "They're not doing anything to help sheep farmers."

Alley used to carry livestock insurance on his animals but it got to be expensive. Each year the farmer has to pay a fee per head of livestock, and in order to collect the insurance he must show the carcass. The problem is that the carcass often disappears or is totally destroyed. Without the carcass the farmer can't collect any insurance.

Howard Morry explains that, in the spring when the coyote pups are small, the female will actually carry off lambs and bring them back to the den, thus making it impossible for farmers to collect carcasses for insurance purposes.

Paul Dunphy admits that it is hard to collect sheep insurance when the carcass disappears but he recommends that farmers carry insurance so they have a better chance of getting compensation.

Alley has seen many of his neighbors and friends cease keeping sheep because of coyote predation and he thinks that the potential is there for coyotes to wipe out the few remaining farmers.

In his experience with the coyotes, Alley has come to learn that they are very smart animals. "They're very intelligent. You've got to try every trick in the book to get him and survive. If you happen to get him in a snare and he gets out, you won't get him again. If you fire at him and miss, you won't get a shot at him again. They're a very, very keen animal," he says.

Talking to Alley one senses that he is fed up with the coyote situation. "I don't know why government is not more interested in this. It's frustrating. The coyote is certainly discouraging farmers and ones already established are going and there's only a few contrary ones like myself sticking in there," Alley says. His frustration is heightened by the fact that there is a high local demand for his lambs. "All my lambs are killed and sold privately. I can't keep up with the demand."

Another sheep farmer who is barely hanging on is Gerald Hewitt of St. Shott's on the southern tip Avalon Peninsula. Hewitt says that he lost about two or three sheep last year, but three years ago he had 19 killed by coyotes out of a herd of 78 animals. "Yes, I've had trouble with coyotes. I have had sheep with the just the throat cut....They weren't eating the sheep, just killing them like they were killing for fun. I'm after losing a good many sheep and lambs to them. I also lost a two and a-half year-old breeding ram." Today, Hewitt only keeps around ten sheep and he says that it just makes no sense to have them. "You just can't afford it. The coyote has had a big impact on the business, I've cut back." Hewitt says that, in the past farmers used to let their sheep go in May and they'd roam for the summer and into the fall. "But the days of letting sheep roam the open country is over now, thanks to the coyote."

Hewitt believes that an economic opportunity is being lost because many people are deterred from rearing sheep now. "The market is extremely good for them. It's no trouble in the fall to sell lamb. My phone is always ringing with people looking for lamb," he says. Hewitt is dismayed that all the open country around his home is not being utilized as it could be for agriculture. "There's a lot of open country and it's a crime nothing is being done with it, no farming, no pastures. Government should be looking to develop the land to do something with it." In fact, he says that there are only a couple of people left with sheep in his area now and several have given up keeping sheep because of the coyotes.

It seems clear that coyotes are having a major impact on the production of sheep on the island. In a province that is desperate for economic opportunities, the sheep potential lost and held back by coyotes is extremely significant.

A part of the coyote-sheep story is that of the Newfoundland sheep. This is a special breed peculiar to the province, similar to the famous Newfoundland pony. Paul Dunphy says: "The Newfoundland sheep is distinctive to the province and we're trying to preserve it.

There's only about 150-200 animals. They're unusual, smaller than regular sheep and some of them have horns. I am very worried about these sheep. Newfoundland sheep are very important to us and there is a lot of reasons to preserve it."

The protection of the Newfoundland sheep is an important issue and integral part of the sheep-coyote puzzle. Parker concludes that there are two basic ways to prevent livestock losses to coyotes; either remove the coyotes, or keep the predator and prey apart. The latter is accomplished by things like fencing, although this is a challenge, because Parker says that coyotes have been known to dig, jump or squeeze through wire mesh with a six inch opening. Electric fences have been used with some success in other jurisdictions such as New Zealand. But they may be too costly for the Newfoundland-scenario. But Parker concludes that "the single most important preventive management option available to the sheep producer is penning sheep at night."

Henry Hilton's research on sheep farms in Maine has parallels to the Newfoundland case. Hilton writes that in Maine, "sheep pastures have generally been poorly maintained and poorly fenced and have provided little security against coyote predation." He goes on to say that coyotes have preyed heavily upon the state's sheep and have decimated some small family flocks: "By 1985 many marginal sheep operations had ceased to operate." Hilton concludes that the "anecdotal evidence suggests that the localized and focused removal of coyotes in winter and spring may reduce predation rates on livestock..."

The province of Newfoundland has recently announced a new coyote hunting license (more on this later) and John Blake says that the "single greatest reason for the establishment of a coyote specific license was to provide for the recreational and economic benefit of hunters while helping mitigate against such things as livestock depredation. A knowledgeable and effective coyote hunter can be a great benefit to a sheep farmer by helping him remove problem animals."

Barry Sabean, director of wildlife for the province of Nova Scotia, says that they too have had a lot of trouble with coyote predation on sheep. "The problem was that when sheep farmers built their fences years ago, they were only concerned with keeping sheep in, not with keeping coyotes out, because we had no coyotes at that time." Sabean says that for many years sheep farmers suffered heavy losses in Nova Scotia and many smaller producers ceased keeping sheep because it was too costly to prevent coyote attacks. Nova Scotia terminated its bounty on coyotes in 1986 and Sabean says that for several years after that, the money was put into compensating sheep producers for losses to coyotes. "Once government felt producers had enough time to adapt, then compensation was discontinued."

In New Brunswick, biologist Cade Libby says that they did have some problems with coyote predation on sheep, but not recently. "The sheep industry is not too big in New Brunswick."

Likewise, Prince Edward Island biologist Randy Dibblee says that they had a primary problem with coyotes predating on sheep in the early 1980's, even though there isn't much sheep production on the island.

Clearly, there are parallels between the cases of Maine, Nova Scotia, and to a lesser extent New Brunswick, PEI, and Newfoundland.

The following chapter deals with the interaction between trappers and coyotes.

...CHAPTER SIX...

Trappers and Coyotes

A part of the coyote story is how the species will be effected by trappers and how the animals will impact on trappers. At first glance, it might seem very straightforward ; trappers catch fur-bearers, therefore they will take coyotes. However, the equation is not that simple. To better understand the relationship between trappers and coyotes several trappers were interviewed.

Eugene Tiller has been trapping for 35 years on Newfoundland's west coast. He is currently Secretary-Treasurer of the Newfoundland and Labrador Trapper's Association. He has trapped everything and anything available. Tiller, like most trappers, has a full-time job and traps part-time in the fall. "I take two weeks off in the fall. The money I make from trapping is a supplement and I use it to buy my outdoor toys," he says.

Tiller says that, like any new species, it takes a while to learn the tricks of a new animal like the coyote. However, the animals are now getting plentiful and most trappers are catching the odd one. Tiller caught five last year, three in snares and two in traps. Tiller says that about four years ago the Trapper's Association brought in a trainer from another province to educate Newfoundland's trappers in the ways of the wily coyote.

"I don't find them harder to catch, but they are harder to hold. I had this one spot where I tied a fox snare onto a little spruce tree every year. I put out the snare and when I went back the tree was gone. The animal had pulled it up out of the ground. Fox snares won't hold a coyote and you need to use a heavier trap than you would for a fox," Tiller says.

George Jennings trapped the first coyotes in Newfoundland and he's been at the fur for forty-odd years now. He was also one of the founding members of the Trapper's Association. He thinks that coyotes are smarter than a fox. "They're cagey. They are harder to catch than a fox," Jennings says. He has also had coyotes break off fox snares, while Tiller has seen snare wire actually chewed off.

A part of the coyote-trapper relationship is that coyote pelts are not as valuable as pelts like fox. Fur-bearer Biologist Mike McGrath says that "coyotes are not worth a heck of a lot, therefore, very few trappers target the animal. The amount of time and effort it takes to trap a coyote is not worth it." Veteran trapper George Jennings agrees with McGrath. He says that it takes a lot of time and traveling to find coyotes and when you consider the costs it is hardly worth it. "The last few years I haven't bothered too much with coyote because you only get about 35-40 dollars a pelt," Jennings says.

Eugene Tiller is of a like mind. He says that coyote pelts are mostly used in the "trim trade." That is, the fur is used as trim around jackets and the hoods of coats. Tiller says that an excellent coyote pelt might fetch $50.00, while a mediocre one will garner about $30.00. George Jennings says that in 2003 a good lynx pelt went for as much as $200.00, while a good fox skin earned about $70. So, the coyote's pelt is not as valuable as the fur of other species.

For example, Newfoundland Fur-Bearer Biologist Ivan Pitcher says that in the 2002-2003 trapping season, there were 96 coyote pelts exported, 699 lynx and 6,147 fox pelts. The fox pelts were broken down into three groups: 4,120 red fox; 411 silver fox and 1,616 crossed fox.

Coyote pelts averaged $43.08, while lynx skins averaged $187.56. Red fox pelts averaged $47.03, while silver fox pelts went for $49.45 and crossed fox pelts garnered $92.44.

Species Management Co-ordinator Wayne Barney cautions that numbers of fox, lynx and coyotes taken in the 2002-2004 may be inflated due to the increased effort under the rabies eradication program.

In 2001-2002, there were 45 coyote pelts exported at an average price of $30.37. That same year saw 7,122 fox pelts taken, and 633 lynx.

Going back to the mid-1990's, from 1993-1996 there were no coyote pelts exported. In 1996–1997 there were three pelts, while in 1997-1998 there were five.

Not only is the coyote pelt less valuable and requires much time and effort to catch, it also requires more work to skin and cure the pelt. So trappers may be less likely to target the species.

Eugene Tiller says the coyote is bigger than a fox and needs more work. "Basically, you use the same drying frame as for a fox, only it's larger. But they are a bit tormenting. They're a little bit harder to skin because the fat and hide stick together. It is more time consuming to skin a coyote. And, they have a strong smell. So strong my wife asked me not to skin them anymore in the basement," Tiller says.

Another aspect of the coyote-trapper relation is that fox populations may actually decline in the face of growing coyote densities. Recall that it has been found that wolves would displace coyotes, which would in turn displace their smaller dog-cousins the fox. Parker writes that "High densities of coyotes tend to limit both the distribution of fox territories and their numbers." But Parker cautions that the actual mechanics of coyote-caused fox declines are not known. So it may happen that as Newfoundland's coyote population increases, the numbers of foxes available for trappers will decrease. This could be significant, because in 2003 there were approximately 7,000 fox pelts shipped out for export. If the average pelt fetched about $50.00, that's $350,000 dollars generated.

Biologist Mike McGrath is aware of the implications of a burgeoning coyote population on other fur-bearers. "We take 700-800 lynx in peak years. Nowhere near the thousands of foxes we harvest. Red fox is really the bread and butter of trapping." Henry Hilton's research on the state of Maine found that "there was concern among some trappers that red fox, fisher and perhaps other highly valued predators would be adversely impacted by emerging coyote populations."

Litvaitis has studied relations between coyotes and other carnivores such as wolves and foxes. He found coyotes dominate foxes and that "foxes and coyotes converged on similar foods more frequently during summer when fruits were available." However, during the winter and spring the coyote's larger body size and ability to hunt in packs allowed them to take big game, namely deer. Litvaitis suggests that the "local distribution and abundance of red foxes may be limited by interference competition with coyotes." This same researcher found that coyotes will co-exist more easily with the highly valuable lynx.

In New Brunswick, Cade Libby wears two hats as both a biologist and a trapper. He says that he has noticed that in some areas there seem to be more coyotes than red fox, "but I haven't heard too much about it."

However, it has been postulated that as coyote densities increase and fox population decrease, other species, like pine marten that are subordinate to foxes, may increase. Mike McGrath says that "in all other places where coyotes expanded their range fox densities dropped. That has ramifications for the survival of many other species. Some places in the northeastern United States where coyotes expanded their range, the fox population dropped, leading to more waterfowl, because fox is an efficient predator on waterfowl. So there will be all kinds of interactions that you can't predict."

However, while coyotes may displace foxes, they may also provide food resources to foxes by leaving the remains of prey, like caribou. For example, Newfoundland and

Labrador Hydro Biologist Brent Sellars has witnessed coyotes feeding on a caribou carcass with several red foxes and two bald eagles nearby, waiting their chance to feed on the remains. Sellars has taken pictures of caribou carcasses picked clean to the bone by a series of animals feeding on it.

Eugene Tiller is also keenly aware of the possibility that fox stocks may plummet as coyote densities increase. "I spoke to a trapper from another province and he told me that when coyotes were first around in his province he might have caught 2 coyote and 50 fox. Ten years later he probably caught an equal number of each."

Tiller says that it would hurt him if the fox crashed and so he is preparing by learning as much as he can about the coyote. "If fox numbers decrease and coyote numbers increase I will spend a bit of time at it (pursuing coyotes)."

Not only may coyotes displace foxes and so hurt trappers that way, they can also be a pest and destroy animals in traps and snares. Eugene Tiller: "Coyotes are giving us more trouble than some people anticipated. I had about three foxes torn up last year by coyotes. It was enough to be a nuisance. The coyote turned a $50.00 fox pelt into a $3.00 pelt." George Jennings has also had foxes destroyed by coyotes. "A coyote will eat a fox in a snare. I have lost several of them."

Another aspect to the trapper-coyote relationship is that the trapper's family may worry more now that there is a little wolf in the woods. Eugene Tiller says that his wife likes for him to carry his .22 caliber rifle because of the coyotes.

Tiller says: "The coyote is here now and we're not getting rid of them. I read somewhere that if anything happened to the earth and all the animals died, the last survivor would be the coyote."

George Jennings is aware that coyotes may have a negative impact on foxes but he thinks the coyote is also doing damage to the valuable beaver stocks. "Coyotes are taking beaver when they come ashore to cut trees. I know areas where the beaver has disappeared."

Clearly, the arrival and expansion of the eastern coyote has meant big changes for Newfoundland's trappers.

The coyote's range expansion in Newfoundland also impacts on resident hunters.

... CHAPTER SEVEN ...

Resident Hunters and Coyotes

In many jurisdictions where the eastern coyote is found there is conflict and competition between the dog-predator and the human predator known as the sportsman. Very often coyotes and sportsmen pursue the same species, such as white-tailed deer, caribou, moose and snowshoe hare.

Coyotes hunt, stalk, chase and kill big game because they must eat. In today's world most sportsmen hunt for reasons other than sustenance, such as the thrill and excitement of the chase, although there are cases where big game meat is still an important dietary supplement. Historically, this has been the case in Newfoundland, where resident hunters, locked into a battle to survive, preyed on native caribou and the introduced moose to feed their families.

Rick Bouzan is past president of the Newfoundland and Labrador Wildlife Federation (NLWF), an umbrella organization that represents rod and gun clubs in the province. It has over thirty-thousand affiliated members.

While time may have lessened the need of most Newfoundland families to kill a moose or caribou for the winter, Bouzan says that there are members of the NLWF who rely on big game as a food source. Bouzan says that he feels betrayed and let down by government. "I blame the bureaucrats more so than the elected politicians. This (coyote) could possibly mean the end of our native caribou and have a major impact on moose." He is unimpressed with the new shooting season announced for coyotes (more on this in a later chapter). "You can't shoot what you can't see. You can't shoot what you don't know how to hunt. We need to educate our hunters about how to hunt coyotes."

Bouzan goes a step further and says that the shooting season for coyotes is totally ludicrous."Only a bureaucratic idiot would think that some Newfoundlander would go and hunt coyotes for sport. The gasoline alone makes it prohibitive, as does the cost of a special rifle. Why would anyone spend money to go shoot a coyote when you can't eat it and the pelt is worthless?"

Veteran hunter and outdoorsman Ed McGrath, of Shea Heights, couldn't agree more with Bouzan. "The cost of the license for coyotes is prohibitive. Then it costs money to get where the coyotes are. You buy a license, leave and go look for a coyote and you may not see one."

Since coyotes are often shot at long distances, the preferred rifles for shooting them are. 22 caliber, center fire, high-velocity rifles. Wildlife manager John Blake admits that .22 center fire rifles are costly. For example, such a rifle, with a telescopic sight, ranges in price from $700-$1000 dollars. Ammunition is about $14.00 for a box of twenty shells. But Blake thinks that coyote hunting will catch on. Although he doesn't know the exact numbers of coyote licenses sold so far, Blake says it is fair to say that "interest has been high."

Retired wildlife officer Bruce Porter is of a like mind. "I think interest will increase in shooting coyotes, even though Newfoundlanders are not into predator hunting. I have a friend in Pennsylvania who tells me that hunting coyotes is a fast growing sport down there. In any case, I agree with the hunt, I recommended it ten years ago when I still worked with the wildlife division."

Neil Coffey is a longtime caribou hunter and he usually hunts the Middle Ridge herd near what is known as Meta

Pond waters. Coffey has seen coyotes on the open, rolling hills of the south coast caribou country. And he has seen the remains of caribou killed by coyotes. "We hunted in the snow last year and every set of caribou tracks we saw had a set of coyote tracks behind it...but I don't know if the coyote is the villain. But I think he is figuring into our wildlife problems."

Coffey's son Dennis actually shot a coyote last year while caribou hunting. "He saw two coyotes and he imitated the howl of a coyote and they came in to about 300 yards. He shot one with his .270." Licensed big game hunters are permitted to shoot coyotes encountered in their designated hunting areas. Neither Coffey nor his two sons have bothered to pick up the new coyote license yet, but Coffey is concerned that moose and caribou will end up being shot out of season by poachers posing as coyote hunters.

Monroe Greening, of Clarenville, believes that the license to shoot coyotes is "useless," since the wolf-dogs are so hard to shoot. Greening has been hunting and trapping in central Newfoundland for 45 years. He is a member of the NLWF and he is quite concerned about coyote predation on caribou, a species he hunts every fall in the Middle Ridge area. "The coyotes are destroying our moose and caribou. Government has to do something or there'll be nothing left."

Greening says that over the past few years he has noticed fewer and fewer caribou calves, and this is consistent with the results of a radio-collar study conducted in 2003. In the study, 30 calves were radio collard in Middle Ridge and all died, many were lost to predation (more on this later). Four were confirmed to be coyote kills, while 15 were confirmed black bear kills.

As a resident hunter, Greening sees the coyote as a competitor for the species he loves to pursue and eat. He has witnessed coyotes chasing caribou over the open barrens and he's found caribou carcasses ripped to shreds. Concludes Greening: "Coyotes are having a major impact on caribou herds and it'll just get worse as their numbers grow."

John Blake is not only a wildlife division employee but he is also an avid hunter. Blake hunts everything from waterfowl to big game, using rifles, muzzle-loaders and compound bows. He admits that coyotes are hard to hunt and that Newfoundlanders are going to have go through a learning curve to successfully hunt them. "People learn to hunt over time from their fathers, uncles, brothers and friends. But there is no peer knowledge of coyote hunting in Newfoundland. The same thing happened when trappers started trapping coyotes, they had no experience with the species. Initially, I think few coyotes will be shot, but as people learn more I think the harvest will go up." Blake, who has bagged several coyotes himself, and who has called out a lot more than he's shot, says that coyote hunting is a challenge like no other in Newfoundland.

Two main methods are used to hunt coyotes. The first one involves baiting the coyote. "In New Brunswick they've been successful in baiting coyotes with road-killed deer or other bait staked into the ground so they can't be dragged away. But black bears are more susceptible to baiting."

A second, more common method of hunting coyotes involves the use of a calling device. Typically, the callers either imitate the challenge call of another coyote, or a mating call. Coyotes are extremely vocal animals and communicate through a series of howls, barks, yips and yaps. This presents a weak spot that hunters can exploit to bag the crafty animal.

Another type of call imitates the sound of a prey species in distress, such as the high pitched squeal of a rabbit. Thoughts of an easy supper usually brings a coyote running if he's in the neighborhood. Calls can be made using either hand held, inexpensive calls or electronic calls. "I called out eight myself this winter past, all of them pairs and shot seven of them. This was on the Gaff Topsails and in the Serpentine Valley," Blake says. (see photo).

But calling them out is only half the battle. Their keen eyesight and sense of smell means that hunters have to be carefully positioned, camouflaged and ready to fire. Hunters usually only get one chance to shoot.

Will coyote hunting take off as a sport in Newfoundland? Will hunters spend the time, money and effort to learn the new skills required to pursue coyotes, or will hunters just kill them incidentally as they encounter them?

Guide and hunter Chad Snow is interested in the shooting season for coyotes. "One of my buddies has a rabbit squealer and we're going to get out and test to see if we can call out some coyotes. I definitely will try to shoot them this fall."

Outfitter Wayne Holloway wears two hats, because he is also a resident hunter. He says that he has shot several coyotes, but he wonders if many hunters are willing to drag a 40 pound animal five miles back to a boat, then transport it back to camp and skin it just for a $25 or $30 dollar pelt?

John Blake says that, while coyotes are having an impact on the island's ecosystem, like all wildlife they deserve our respect and deserve to be harvested humanely. Blake is concerned that the creation of any negative image in the eyes of non-hunters will hurt hunters much more than coyote predation on big game. "We cannot allow ourselves to portray an image of wanton destruction to the non-hunting public," Blake says.

Trapper Eugene Tiller is also a resident hunter and he has gone so far as to have an electronic caller shipped in so that he can learn to successfully hunt coyotes if the population of red fox crashes. But he says that having a coyote season in spring and summer is useless, since the pelt is no good until the fall, when the weather gets colder and the animal's coat thickens up.

Gord Follett, editor of The Newfoundland Sportsman, has gotten many reports of coyotes killing caribou. He likes the idea of the coyote license.

Follett thinks the new hunting season is a great idea. But he is surprised there hasn't been more of a reaction from what he calls the "coyote huggers" who have opposed his columns about shooting coyotes.

"I just wish coyotes never arrived in Newfoundland, or

that they were shot and killed as soon as the first ones were spotted. Forget this 'part of nature' or 'food chain' crap. We don't need them. We don't have snakes or alligators in this province either and life has been fine without them."

Follett says that he is concerned that his prize beagles might fall prey to coyotes when he takes them rabbit hunting in the fall. "If the dogs are gone for a half hour and I can't hear them barking, or hear the bells on their collars, I worry about them. I wonder if they got caught in a snare, or if a coyote has taken them." While Follett hasn't actually hunted coyotes yet, he definitely plans to go. "I won't just go moose hunting or rabbit hunting and hope to bag a coyote, I'll target the coyote."

The outdoors writer has been reading up on the coyote in an effort to learn as much as he can about this predator. Follett says that from what he understands, shooting a coyote is a much bigger challenge than getting a moose or caribou. He thinks the fact that coyotes are so hard to shoot will influence hunter's interest in the sport.

"You might get a fair amount of interest in the next couple of years, but I think it will gradually die off as people realize how difficult it is to hunt. I think that's too bad because I'd like to see more people get and stay involved in coyote hunting. But you'll get really serious hunters go after coyotes and then moose and caribou hunters might fluke into one occasionally."

Follett says that he has been getting a lot of calls and e-mails about what type of gun to use for coyotes, more support for the notion that Newfoundland hunters have a significant learning curve to go through for hunting coyotes. He says that there are many videos available on how to call coyotes into shooting range.

"From what I've seen and heard, it is truly a challenging sport; one of, if not the most challenging hunts available in Newfoundland and Labrador."

But it remains to be seen if coyote shooting will catch on with Newfoundlanders. Will hunters who historically have hunted moose and caribou for meat, shift their attention to this fur-bearing predator? Will resident hunters be willing to invest in special small-caliber, high velocity center-fire rifles just to bag a coyote? Time will tell how the coyote hunt will develop.

John Blake believes that predator hunting is a legitimate activity that, if conducted properly, can provide not just recreation but economic benefit as well. "All across North America, people harvest canids and other fur species by hunting rather than trapping. Using the right firearm, ammunition and hunting techniques, there is no difference in the price of fur received at auction of those harvested with a firearm versus those trapped. I, for one, would prefer to hunt rather than trap fur and I see no reason why this predator hunt cannot be expanded to include fox, within the limits of when the fur is prime."

By way of comparison, the province of Nova Scotia permits hunters to shoot coyotes year round and there is no bag limit. The Director of Wildlife for Nova Scotia, Barry Sabean, says that "it's fair to say that some people are still concerned about the relationship between white-tailed deer and coyotes." Sabean thinks that most hunters in Nova Scotia have come to accept that the coyote is there to stay and will prey on deer, affecting the dynamics of the herd. Hunters are also concerned about small game such as rabbits and when the rabbit population drops people often point the finger of blame at the coyote.

New Brunswick and Prince Edward Island also have generous coyote seasons. In New Brunswick, trappers can take coyotes as fur-bearers, but the species is also classified as vermin, and holders of any license, such as black bear or white-tailed deer, can harvest coyotes they may encounter. Coyotes are also designated a nuisance species and private land-owners can destroy them. Like Nova Scotia, there are no bag limits, although, like Newfoundland, there are certain restrictions on the type of firearm that may be used to harvest coyotes.

In Prince Edward Island coyotes are designated as both game animals and fur-bearers. Hunting is from October 1 to March 31, while trapping is permitted from November 1 through January 15. The province also has an active problem control program. Biologist Randy Dibblee says that he has never called coyotes "vermin." Dibblee stresses that we must have respect for all wild creatures, coyotes included.

However, some of Newfoundland's big game outfitters are hopping mad about the impact coyotes are having on the lucrative woodland caribou herds.

... CHAPTER EIGHT ...

Outfitters, Coyotes and Caribou

The outfitting industry has a long history in Newfoundland. For years non-resident sportsman have paid good money to come to the island to hunt and fish for trophy caribou and prized salmon. For example, after the railway was pushed across the island in the late 1800's, the Reid Newfoundland Company promoted caribou hunting along the track. For years famous sportsman Lee Wulff worked with the government promoting hunting and fishing in Newfoundland and Labrador.

Moose, which were introduced to Newfoundland, and the native woodland caribou, are important species for the outfitting industry. Newfoundland is one of the few places in the world where sportsmen can pursue woodland caribou. Woodland caribou are one of 27 species required for the North American Grand Slam of hunting. Sportsmen are willing to fork over big money for a week long hunting trip and the chance to bag a heavy-antlered stag.

A guided week's hunt at a lodge goes for about $7,000, a significant sum. The net economic benefit of big game outfitting on the island is estimated at $31 million annually. In all there were 291 licenced outfitters in the province in 2002. On the island, 131 companies operate 221 lodges, 98 of which specialize solely in hunting. As for employment, outfitting employs a significant number of people on a seasonal basis. According to the Department of Tourism, approximately 1,100 people work as big game guides during the fall hunting season on the island.

Eric Patey has been running an outfitting operation on the Northern Peninsula for almost twenty years. His guests hunt moose and caribou in big game management zones 3, 39 and 69. Patey is very concerned about the impact the coyote is having on big game stocks in his region, especially caribou.

Patey says that he first became aware of coyote predation on caribou in the River of Ponds area about eight to ten years ago. "That winter about 35 caribou came out near what we call the head of the pond. Every time I went in to the head of the pond on skidoo there was one less caribou. By the spring I think only three caribou survived."

Patey actually witnessed a lone coyote attack a caribou one night just before dark. Patey had gotten quite concerned about the caribou being killed in the area and so he began to make regular visits to the small herd.

"One evening I was watching the caribou when suddenly they took off running. Out of nowhere came a coyote running full tilt across the barrens. He caught hold to one of the caribou by the hind leg and he was so intent on what he was doing he never noticed me on my skidoo."

Patey spends a lot of time skidooing during the winter. He says that he is seeing an "unbelievable number of caribou kills."

Wildlife Officer Bill Green provides evidence which supports Patey's assertion. Green recently spent much time flying one kilometer lines over the Northern Peninsula as part of the province's rabies eradication effort. "We were flying low and slow so we didn't see many coyotes because they'd hear you coming. But on every big waterway on the Northern Peninsula you'd see dead caribou."

It seems that the coyote drives the caribou onto the ice where the animal has better traction than hoofs afford caribou. This is apparently why so many caribou kill sites are seen on ponds and lakes. John Blake thinks that one reason so many kill sites are on frozen ponds is because that's where people are skidooing and thus find the kills.

Newfoundland and Labrador Hydro Biologist Brent Sellars has also witnessed caribou being killed on the ponds and lakes that surround the Granite Canal hydro project. "Coyote are taking down full grown adult caribou in a big way." Environmental consultant Bruce Porter, who flies with Sellars, says that he has seen a number of kills on Granite Lake. "Coyotes are taking down a number of caribou. They are not selective and they aren't just taking calves. They are having an impact."

Eric Patey often locates the ravaged carcasses of caribou by the crows flying around overhead. Frequently, Patey finds that the coyotes have just chewed the hind hams off the caribou and left the rest of the carcass. Patey describes the coyote as a very wasteful animal. "It kills, leaves and goes on." Similarly, biologist Wayne Barney also notes that coyotes seem to eat the cheeks and tongue of caribou and the meaty hind haunches.

In March 2004, Eric Patey was skidooing one day when he came across a sight he can't forget. An adult bull moose had been driven into deep snow and attacked by coyotes from the rear. "It had the hinds chewed off and it was still alive."

The fact that coyotes are taking moose comes as no surprise to Gary Tuff of Gander. Mr. Tuff was driving along the Trans-Canada Highway on the morning of June 30, 2004 when, just east of Port Blandford, he witnessed a coyote chase a moose out of the woods and across the highway. The moose escaped, but Tuff was able to snap three photos of the coyote lurking about the roadside brush.

Biologist Kim Bridger has been studying the stomach contents of coyotes collected from all over the island. Her research confirms that coyotes are killing both moose and caribou. "I am finding large chunks of flesh in their stomachs. Coyotes are preying on moose, too. Most meat found in the stomachs has hair and bone attached, therefore it's easy to identify as moose or caribou." Bridger explains that she was surprised to find that snowshoe hares weren't playing a bigger role in the diet of coyotes. "Maybe there are so many caribou and no other predator so the coyote is not so linked to the snowshoe hare in Newfoundland."

In fact, Gerry Parker's book about the eastern coyote theorized that snowshoe hare would be the primary food for coyotes in winter. But based on the results to date from Bridger's study that does not seem to be the case.

Wildlife manager John Blake says that the thing about coyotes is that in much of their range expansion, white-tailed deer has been the main prey of coyotes. "Here in Newfoundland it appears to be caribou. There's definitely dynamics going on and we know that they are killing moose to a lesser degree," Blake says.

Eric Patey is angered to think that moose and caribou are being destroyed, robbing him of economic opportunities, and nobody seems to be doing anything about it. He would like to see marksmen sent in to kill coyotes and protect the caribou calving grounds. He says that nothing has been done by government since the coyotes were first spotted in the mid-1980s and he thinks now it is too late. "I remember years back when people were starving and the moose and caribou populations were kept down by poaching. The herds came up when people stopped poaching. Moose and caribou are too valuable to let the coyote go unchecked."

Patey says that the coyote is effecting his business. He explains that every client (i.e. non-resident big game hunter) he brings in creates a week's work for his guides. "Last year my business generated 250 work weeks. I employed 32 people last year. Some got as much as ten weeks work, but the average was five to seven. For a lot of men who are fishermen, guiding in the fall is important. If you add in what outfitting brings to Newfoundland it's important."

Patey emphasizes that he has traveled to sport-shows promoting Newfoundland's woodland caribou hunting. "It's one of the few places you can hunt woodland caribou."

Patey is concerned that the caribou herds are being decimated and that outfitting businesses like his will be hurt. He predicts that in two years he will lose his caribou licenses which will hurt the local economy. He is frustrated that no-one in government seems accountable for what he sees as devastation of the caribou herds by coyotes. "I don't think I know what coyotes do, I know what they do. I see it when I am skidooing. Caribou can't be born fast enough to replace what's being killed. One coyote can kill a caribou like a man could take candy from a child."

Chad Snow is a licensed big game guide who takes non-resident sportsmen hunting for moose and caribou near Lewisporte (areas 22, 22A and 68). He has witnessed coyotes chasing caribou and, like Eric Patey, Snow is very concerned about the coyote's impact on caribou and the potential spin-off on his livelihood. "I have seen a big difference in caribou in the last year. I saw a lot less caribou and I only saw two calves. Also, the caribou seemed to be staying in the green woods more. I think that's all related to the coyote."

Snow thinks that the caribou is in such bad shape that the herds might soon be closed to hunting, which will cost him six weeks of important guiding work in the fall.

Wayne Holloway is another big game outfitter who runs five camps in the Middle Ridge (area 64) with his brother. Holloway is extremely worried about the impact of coyote predation on caribou. Like Eric Patey and Chad Snow, he thinks the future looks bleak for Newfoundland's woodland caribou. In fact, he thinks woodland caribou are headed for the threatened species list.

Holloway says that he and his guides first began noticing significant evidence of coyotes about five years ago. But they didn't see an impact on caribou at that time. As time went on the evidence become more voluminous. "Coyote predation on caribou is a problem regardless of what the government says. There is evidence that adult caribou are being killed by coyotes, so wouldn't calves be even more susceptible to this predator?"

Holloway says that even though he has not had his caribou license quota reduced yet, the predation of coyotes on Middle Ridge caribou is effecting his business in that there is less available game; "It's only going to get worse."

Biologist Mike McGrath admits that hunter success rates on the Middle Ridge herd are declining and calf recruitment is very low. In fact, McGrath says that things are so bad he has stopped hunting in that area.

Last year the Wildlife Division radio collard 57 caribou calves in three different herds (Middle Ridge, Mount Peyton and Gaff Topsails) and tracked their movements. When the transmitters stopped moving, scientists moved in and examined what happened to the animal. Big game biologist Shane Mahoney is the chief of scientific research for the province. He explains the particulars of the radio-collar study.

The three caribou herds were chosen for study because of their different ecological circumstances. The logic behind the study is that the percentage of collared calves that survive is a gauge of how many calves are being added to the herd.

"We put 17 collars on the Gaff Topsails and one calf survived. We put 30 radio-collars on the Middle Ridge herd and none survived, and we put 10 radio collars on Mount Peyton and 3 survived, although one has since died. Out of a total of 57 collars, only 3 caribou are now still alive. No caribou, no animal population can sustain itself with that level," Mahoney says.

That means that something like 93 percent of the radio collard calves were lost to predators.

Mahoney states that outfitters know the caribou herds are declining and the radio collar study was an attempt to quantify and assess why "our caribou populations are dropping like a stone."

Mahoney explains that the cause of death varied greatly from herd to herd and that the coyote was not the sole predator killing juvenile caribou. In the Gaff Topsails study, 4 of the 16 calves lost, or 25 percent, were confirmed to be coyote kills. Therefore, the assumption is that coyotes are killing 25 percent of the calves in this herd. However, Mahoney cautions that this 25 percent figure could be even higher because there were a few kills in which the biologists were unsure what killed the calf.

In the Middle Ridge herd, all 30 radio collard calves were lost. But coyotes only killed a confirmed number of 4 or 13 percent. Black bears on the other hand killed 15 of the 30, 50 percent. Again, Mahoney cautions that the figure for coyote predation could be higher.

In the Mount Peyton herd, where 10 calves were collared and 7 were lost, coyotes did not kill any of the calves, while lynx and black bears did.

The results of this one year study show that in the case of juvenile caribou, coyotes have shown themselves to be a significant new predator and appear to be killing caribou in significant numbers in addition to predation by lynx and black bears. This is significant because it seems that coyotes, lynx and bears are not just sharing the percentage of calves that lynx and bears used to take, but the coyote is taking additional calves. Mahoney says that biologists hope to repeat the study next year. He admits: "The rate of mortality in our caribou herds is frighteningly high."

The radio collar study shows caribou calves are falling like dominoes, but not only to coyotes.

There are many reliable reports of coyotes killing adult caribou, but biologists are unsure of at what percentage. "If one coyote is able to kill an adult caribou then a reasonable number of coyotes could do a lot of damage to the herds," Mahoney says. Similarly, John Blake agrees that coyote predation on adult caribou is a significant issue.

Mahoney cautions that while some people will inevitably blame the caribou crash on coyotes, there may be other factors at work. For example, caribou body size has been getting smaller over the past 10 years. This is caused by lack of food. It seems that the caribou may have over-browsed their range, damaging their food supply, and this is contributing to the population dynamics. But no doubt about it, coyotes are having a major impact on caribou.

Mahoney: "Let me make it very clear, I wouldn't underestimate the impact that coyotes are having on caribou herds. I think it's a very significant impact but it's not the only factor."

Biologist Mike McGrath cautions that the Middle Ridge herd was stable at about 3,000 animals in the 1970s, by the mid-1980s the herd increased to about 22,000 or more. It peaked and is now coming down again. McGrath says that caribou herds are never stable, they are in a state of flux. "The densities of caribou will increase, the habitat will degrade somewhat, calf survival decreases and they're more vulnerable to predation because they're not as healthy. So there's all kinds of interactions that can come into play."

Shane Mahoney also points out that in the early twentieth century caribou populations on the island crashed heavily. "In 1900, I estimate we had 80,000-100,00 caribou. It's why the great sportsmen came here and marveled at our caribou herds. By about 1920 all hunting was closed. (The herds crashed). The wolf went extinct about 1922. People don't appreciate history."

Across the Gulf in Nova Scotia, white-tailed deer populations plummeted as the coyote population peaked. But that province's director of wildlife, Barry Sabean, says that Nova Scotia had deer crashes before the arrival of the coyote. "But the coyote made the population crash faster and drove it lower and slowed the recovery of deer. The deer did recover and are now crashing again."

The Nova Scotia deer example does not ease outfitter Wayne Holloway's concerns over caribou. Holloway is very worried over the results of the Newfoundland radio-collar study. He says that it is known that there is zero percent recruitment or 100 percent calf mortality for many herds. And, they're killing adults in the winter. "So the breeders are not making it through the winter and you're calves are not making it to October and you've got hunting on top of that. Where does that bring you?"

License allocations for resident caribou hunters have been reduced in area 64-Middle Ridge over the past few years and resident hunter success rates have been dropping. So a reduction in quota for non-residents, that is outfitters' clients, may be just around the corner.

Holloway: "The problem is that the coyote pelt is valueless and the prey they're after contributes $7,000 per unit to our economy. If a cattle rancher with 150,000 head of cattle discovers 5,000 mountain lions on his range, would he sit back and let the universe unfold or would he intercede?" Holloway estimates that in the last five years a half billion dollars in moose and caribou has been lost to coyote predation. "These species, like moose and caribou, are our most precious per unit renewable resource in Newfoundland."

Outfitters feel so concerned about coyote predation that the Newfoundland and Labrador Outfitters Association spear-headed a proposal for a coyote population containment program. This included a petition and was submitted to the provincial government in April, 2003. Other groups involved with the petition included the Newfoundland an Labrador Wildlife Federation, and the Newfoundland Sheep Breeders Association, and the Rural Rights and Boat Owners Association. The proposal included an open season on coyotes year-round, education programs on hunting the coyote, introduction of a bounty system and allowing the use of poison. To date very few of these recommendations have been implemented.

Biologist Mike McGrath admits that there is an issue with coyote predation on caribou. "The thing is, black bear predation was always happening, so coyote predation was in addition to it, and new. Therefore, this could be contributing to changes in population dynamics of caribou."

McGrath says that they plan to repeat the caribou calf radio-collar study this year, but he cautions that caribou calf survival and coyotes are not linked in a straightforward manner. "It's not as simple to say eliminate all coyotes then you're not going to have a caribou survival problem. For example, we went through a massive increase in lynx numbers in recent years. This predator had a major impact on caribou in the 1970s." McGrath says that black bears are the number one predator of caribou calves and that the biologists don't really know what's going on with that. Bear densities may be increasing.

A problem facing big game biologists as they attempt to unlock the puzzle of caribou-coyote relationships is that the two species only co-exist in a couple of places; Alaska and the Gaspe` Peninsula of Quebec. In most of its range, the coyote's primary prey has been white-tailed deer. However, there are no white-tailed deer in Newfoundland and there is a scarcity of research on caribou-coyote interaction.

For example, in Alaska winters are similar to those in Newfoundland. However, Alaska has two important species that Newfoundland lacks; namely white-tailed deer and wolves. In this province, there are no white-tailed deer for coyotes to prey on, nor are there any wolves to either keep coyote numbers in check, or act as primary predators and leave the remains of moose for coyotes to scavenge. Yet despite this crucial ecological difference, Parker theorized that direct predation by coyotes on moose and caribou during Newfoundland's winter was unlikely, although juveniles, crippled and diseased adults may fall prey to coyotes. However, given what is known today, Parker's ideas on coyote predation underestimated the coyote's ability to adapt and bring down large game in deep snow, or on icy ponds and lakes. We know that caribou are a preferred food of coyotes, and that coyotes are killing both juveniles and apparently healthy adults. Parker says that coyotes prey on white-tailed deer that are representative of the general population and are not old, weak or diseased animals. The same seems to be true of the coyote's predation on adult caribou in Newfoundland. It may well be that Newfoundland's wildlife managers also underestimated the potential impact of coyotes on the island's big game herds. This may help explain why the provincial government apparently invested very little in coyote analysis from 1985-2000.

One of the few places in the world where the range of coyotes and woodland caribou have over-lapped is on the Gaspe` Peninsula of Quebec. In that location ranges a small group of approximately 250 woodland caribou, the last herd remaining south of the St. Lawrence river. The caribou live at elevation greater than 700 meters in Parc de la Gaspe`sie and are protected by Federal Species at Risk Legislation. Parker says that in 1987 biologists became alarmed about the herd's future when it was learned that while most females were giving birth, very few calves were surviving their first winter.

In 1989 and 1990, 25 caribou calves were radio collard and 16, 64 percent, died the first summer. Of the 16, 7 were killed by coyotes, 3 by black bears and 1 by a golden eagle, while for 5 of the calves the precise cause of death could not be determined. In the early 1990's, coyotes and black bears were removed from the park through a trapping program and caribou calf survival increased. But Parker says that biologists were reluctant to attribute increased calf survival solely to predator removal, but they did admit that removal of predators reduced caribou calf mortality.

Compare the death rate of the 1989-90 Gaspe` radio-collar study (64 percent) with the mortality rate in the recent Newfoundland study (93 percent) and it is quickly seen that caribou seem to be in real trouble on the island.

The trapping season on coyotes in Parc de la Gaspe`sie was maintained from 1990 - 1996 to help improve caribou calf survival rates. There may be lessons in this for the Newfoundland case, given the extremely high rate of calf mortality discovered in the radio-collar study.

Long-time journalist Bill Callahan served as a cabinet minister in the Smallwood administration. Callahan told me that he recalls in the 1950s and 1960s, when it was determined that lynx were playing a major role in keeping caribou numbers down, the provincial government introduced an intensive trapping program to reduce lynx numbers and help stimulate caribou calf survival. Callahan believes that the current provincial government needs to take similar action to deal with the apparently serious problem of coyote predation on caribou.

... CHAPTER NINE ...

Coyote Management in Newfoundland

As we know, coyotes are having a significant impact on various resource sectors like sheep farmers, trappers and outfitters. One of the people responsible for managing Newfoundland's growing coyote population is Species Management Co-Ordinator Wayne Barney.

Barney says that there are four categories of coyote management in Newfoundland and Labrador. First, the introduction in 1989 of a trapping season from October 20-February 1.

A second management option was created in the early 1990's to deal with coyotes on a problem basis. This was a permit to allow people such as livestock producers to remove troublesome coyotes.

In 2002, a third management option came into being when coyotes could be shot by any holder of a valid hunting license. This allowed a moose hunter in area 22, for example, to shoot a coyote if seen while legally hunting in that area. This was an incidental harvest technique allowing anyone with legal access to a firearm and a valid hunting license to harvest coyotes.

Recently, the fourth category of coyote management was introduced for coyotes when a coyote shooting season was announced in 2004. In this management option, coyotes have been added to the legal definition of game in the province. Hunters can now buy a specific license at a cost of $10.00 dollars to hunt them for ten months of the year, from the second Saturday in September through the second Saturday in July. Although in its first year, it initially ran May 1, 2004 to July 9, 2004.

According to an information pamphlet, "Coyote," prepared by John Blake, the rational behind the coyote hunt is to provide "enhanced recreational benefit to the province's hunters while helping to control livestock and big game predation."

The coyote license is intended for the serious hunter who specifically wishes to pursue coyotes, rather than just taking one as a result of a chance encounter while moose or caribou hunting.

There is no bag limit to the coyote license but there are firearm restrictions. The regulations state "the holder of a coyote license may hunt, take or kill coyote with a center-fire rifle not greater than .225 caliber or a shotgun using shot size #2 or larger." It is thus illegal to hunt coyotes with the popular rimfire .22 caliber rifle. According to Blake, rimfire .22's simply do not have the energy to consistently and effectively put down coyotes that can weigh as much as 40 pounds or more.

Another aspect of the coyote management plan is that the Inland Fish and Wildlife Division is continuing to collect coyote carcasses and for the year 2004-2005 will pay a fee of $25.00 for each carcass submitted for study.

Based on statistics kept from furs exported out of the province, from 1993 to 1996, there were no coyote pelts exported out of the province for sale. But fur-bearer biologist Ivan Pitcher explains that these numbers may not accurately represent the number of coyotes taken because people may have kept the pelts because of the novelty of the species.

In 1996-97, there were three coyote pelts exported, and in 1997-98 this number increased to five. In 2001-2002 there were 45 coyote pelts exported, while in 2002-2003 there were 96 coyote pelts shipped. So, there is a clear trend of increasing coyote harvests and this is consistent with the coyote completing its range expansion over the island and beginning to increase its densities.

The numbers for 2004 are not yet fully collected but Wayne Barney thinks that it might well be over 300. "This is consistent with what we expected. First the coyotes spread out across the island in low densities and now they are starting to increase in number."

John Blake says the province's coyote management plan has two objectives. First, to mitigate coyote problems, such as those around sheep farms. Second, to provide a recreational opportunity for hunters.

However, outfitter Wayne Holloway is not impressed with the management plans in place for dealing with coyotes. When asked what he thinks of letting licensed big game hunters shoot coyotes, Holloway says that this is "totally useless." Holloway says that a hunter pursuing a moose isn't likely to stop and kill a coyote, given that it might interfere with him killing a moose. Also, once the moose is bagged, the big game license is not valid any longer, so that hunter is taken out of the field and therefore cannot hunt coyotes. Plus, Holloway says, most rifles your average Newfoundlander would have for moose hunting, such as a .303, are not well suited to shooting coyotes. "The kind of rifles you need for shooting a coyote are flat-shooting, high velocity, low caliber center-fire rifles."

Holloway says that coyotes are as prolific as rabbits, but if you "knock off" 10,000 females in March then they won't be alive to breed. "And, you deal with it every year." The new hunting season will allow hunters to prowl the country on skidoos during the winter.

Finally, Holloway believes that the present management plan is really just a whitewash. "All they're doing is covering themselves and making it look like they are concerned. They should have been concerned 18 years ago when the problem was fixable...At the time, in 1985, it was spring and snow was everywhere. With some effort a half-dozen coyotes could have been rounded up and done in. But what did they do? They dispatched a wildlife officer and he flew around for an hour and he didn't see the coyotes and he came home and that was the end of it."

Fellow outfitter Eric Patey concurs. "It's incredible that government left coyotes unchecked for such a long period of time. The policy is only a joke to appease people, it's not a real genuine plan. It's too little too late and it's not going far enough."

Government officials say that a bounty on coyotes is not being considered for Newfoundland and Labrador. A bounty essentially means that government pays a fee for every coyote pelt turned in during a year-round open season.

Wayne Barney thinks bounties don't work. "Bounties, where they have been implemented across North America, have not worked. People realized it was a big waste of money. For example, Nova Scotia, at the peak of its bounty, was harvesting 8-900 coyotes. But no jurisdiction has rid itself of coyotes. So why chase a mechanism we know won't work? We certainly won't eliminate coyotes."

But at present Newfoundland is offering a cash payment of $25.00 for coyote carcasses turned in by trappers. Wayne Barney explains that this reward is also used for lynx and helps gather scientific information like age/sex ratios, diets, diseases, reproduction, etc. The distinction between a bounty and the cash payment offered by government seems very fuzzy at best. On the one hand, government officials say that paying people a fee to facilitate coyote capture does not work, yet at the same time they are offering a $25.00 cash reward for coyote carcasses turned in.

New Brunswick Biologist Cade Libby says that bounties don't work, but outfitter Eric Patey disagrees with the way coyotes are being managed. He thinks the shooting season for coyotes without a bounty is a joke. Patey says that it is slow in rural Newfoundland during the winter and a lot of men would go after coyotes if there was some money in it (i.e. if there was a bounty). "In winter if people were sure they'd get their gas paid for and get the bounty they'd be in after the coyotes."

Similarly, Rick Bouzan, of the Newfoundland Wildlife Federation thinks there should be a bounty. "We need to go to war with coyotes and implement whatever means are necessary to knock back the numbers. This could possibly mean the end of our native caribou. This is like seals tearing the guts out of codfish."

Biologist Mike McGrath knows that there are recommendations coming from individuals and groups to try and exterminate or eradicate these animals. "But that's not a feasible option. Attempts have been done elsewhere. Right across North America there have been many, many attempts to eradicate the coyote but without any success anywhere I know of... The coyote will certainly make its own room. The jury is still out on their impact on caribou and arctic hare."

Undoubtedly, the experience elsewhere indicates that bounties are not successful coyote management tools. Parker and Moore suggest that "it is illogical and impractical to advocate widespread control or elimination of the coyote given the growing non-user interest in wildlife populations and the long history of futility in control attempts through the past 65 years." In the state of Michigan almost two million dollars bounty was paid out for 111,569 coyotes taken between 1935 and 1970. Parker says that "after 35 years of continuous bounties on coyotes, only five fewer were killed in 1970 (3,021) than in 1935 (3,026)."

Several Canadian provinces have had short-lived bounty programs. For example, starting in 1967, Quebec had a bounty of $35 dollars on all coyotes. This program was discontinued in 1972.

Similarly, Nova Scotia implemented a $50.00 coyote bounty in 1982, just five years after the canid appeared in the province. However, Parker says the program was ineffective in limiting coyote population growth and the program was ended in 1986. That money was then funneled into compensating sheep farmers for coyote damages.

Alberta had a coyote bounty from 1920-1948 and researcher Michael Dorrance says this failed to eradicate coyotes. In 1951, coyotes were designated pests in Alberta, but by 1975 they were given the status of nuisance. From 1949 - 1972 coyote control in Alberta was characterized by the use of poisons. That, too, has been suggested for Newfoundland.

For example, Wildlife Federation member and avid hunter, Monroe Greening, says that back in the 1950s and '60s when it was thought lynx were causing high caribou calf mortality, a poison program was introduced. "It (poison) is one alternative, even if we have to do it as a last resort. I don't want to see it, but if we've got to use poison, do it."

Sheep farmer Wade Alley says that in some provinces if a farmer has a problem coyote, a specialist comes in and sets out poison. "But our wildlife division won't even think about it. There's no help."

The proposal submitted to government in April 2003 by the Outfitters Association, the NLWF and the sheep producers also recommended the use of poison to control coyotes. Species manager Wayne Barney says that the use of poison to control coyotes is not being considered. "For one, it is against the policy. Second, poisons on the scale needed to address coyote elimination is of such scale the impacts would be felt throughout all wildlife populations, not just the target species." Barney also explains that if the goal was control of populations, poisons would have to be used indefinitely, as coyotes would keep reproducing.

Biologist Shane Mahoney dismisses poison as a coyote management tool. "The problems of a poison program are well known. How do you deliver the poison in a way that it doesn't effect non-target species? The environmental assessment policy would have to come into play. Who would pay for it? Millions of coyotes have been taken out by control measures but coyote predation has never been eliminated."

Mahoney points out that the radio collar study on caribou calves shows that bears and lynx are also taking calves, not just coyotes. So a poison program aimed at coyotes, even if it was successful, would not automatically lead to increased caribou herds.

Mike McGrath says that the coyote is here to stay and it remains to be seen what kinds of population densities will emerge. "If they are causing a problem with caribou when their numbers are low, it might be a big issue later on when the populations of coyotes increase."

In the state of Maine researchers found that managers of white-tailed deer herds had to adequately estimate coyote predation when managing deer. Gerald Lavigne's research found that "because nearly half the deer killed by coyotes were mature deer, winter predation by coyotes may affect herd dynamics by reducing the density of breeding-age does." Therefore, deer managers in Maine had to factor in variations in predation when setting harvest levels. "A failure to consider predation losses could lead to undesirable herd declines."

Wayne Barney says that the process of managing the coyote is evolving since the animal introduced itself. "Now that it's here we're responsible to manage the species."

Connolly's research suggests that while coyote attacks on humans are rare, "more coyote/human interactions can be expected in eastern North America as coyote and human populations continue to increase."

Likewise there is mounting evidence to suggest that Newfoundlanders had better get used to co-existing with coyotes.

For example, Kelley Symonds, of Cottrell's Cove on the northeast coast, saw a coyote alongside the highway just before dark one day. The critter seemed very relaxed and Kelley was able to stop the car and take several pictures of the animal in various poses. She even got out of her vehicle to snap pictures and the animal seemed quite relaxed and posed for photos.

Gary Tuff, of Gander, saw a coyote chasing a moose along the Trans-Canada Highway near Port Blandford one day and was able to stop his car and snap three photos of the little wolf.

Guide Chad Snow says that he recently saw a coyote on a woods road not far from Lewisporte that was at very close range of about 100 feet and it just stood there and stared at him. Also, a coyote was seen dragging off a bag of garbage in Lewisporte during the winter.

Mrs. Jean Kennedy of Mobile, on the southern Avalon Peninsula, was walking her dog along the old rail-bed one morning in February when she came to a bridge crossing a frozen brook. There on the ice-covered stream was a coyote out for a stroll. Mrs. Kennedy says that she took out a whistle which she always carries, and blew it, but the coyote ignored her and just kept ambling along the river. "I was pretty close to it, just two or three house lengths away and we were close to the community, you could see the houses from where I was. I was nervous for a while after that and I don't let my dog go into tucks of woods anymore."

Retired wildlife technician Bruce Porter has a reliable report that coyotes have been seen on Fogo Island, marking the colonization of yet another island. He also had a close encounter with a coyote one day in the Granite Lake area. "The chopper dropped me off on a grassy point on a small pond. The chopper lifted off and was going to come back for me later. I was fiddling around, when I noticed there was a coyote looking right at me. It was downwind of me, so it had to smell me, but it was like a big dog, it just ignored me. I threw a rock at it, which missed and it just stayed there. I threw another rock, missed again, but it trotted off into the woods. It was one of the few times I didn't have a camera with me."

Likewise, Brent Sellars had an encounter with coyotes that he won't soon forget. "It was after work and I was fishing on a small pond near Granite Lake one night after work. I heard coyotes howling about a kilometer away for two or three minutes. Then I heard vicious snipping and snapping. Then it got real quiet. I assumed they killed something. This was after work in the evening and it was getting dark. I was alone in the middle of nowhere and not used to the sound of howling. It was the eeriest feeling."

Mike Tulk of St. John's had a close encounter with a pack of coyotes that he won't soon forget. Tulk was goose hunting near Soldier's Pond one evening in late September 2004. Tulk stayed at his goose blind until dark but saw no birds, so he left to walk the half-hour or so to his pick-up on the TCH.

As he was crossing a bog, he heard something running ahead of him; squish, squish squish. When he shone his flashlight around he was looking into the yellow eyes of a large coyote. The wild dog was close, about thirty feet, when suddenly Mike realized there was more than one coyote. It all happened quickly but the dogs began to snarl and growl at him, so he fired a shot to scare them off. However, after walking a short distance he realized that the coyotes were trailing him. He again fired a shot, but to no avail. The coyotes kept tracking him.

"I figure there were at least four or five coyotes. I'd shine the light around and see one here, one over there, another there," Tulk says.

The seasoned outdoorsman had seen coyotes in the area on previous occasions, but usually he caught just quick glimpses as the animals headed for cover. This time, because the coyotes were so persistent, Tulk was concerned about getting to the safety of his truck.

"I was carrying a shotgun and only for that, they would have been right onto me. I wanted to talk about this because if it was a berry-picker or a trouter they could have been in trouble. I wouldn't want to meet them again like that," Tulk says.

Mike Tulk's story is chilling. What might have happened that night if he had been returning from trouting and was only armed with a fishing pole? Will there be more close encounters like this one in the Newfoundland outdoors? Do outdoor enthusiasts such as hunters, trouters and berry pickers need to take precautions as they venture out into the woods?

Clearly, coyotes have spread across Newfoundland and the population densities are increasing. More sightings and interactions with humans are inevitable, given our high outdoor participation rates in this province. People had better get used to the wolf-like howl of coyotes, or the occasional chance encounter as they hunt, fish, hike and berry pick in the Newfoundland outdoors.

Of course, coyotes have been seen in urban areas, such as the one in Lewisporte this past winter, or one spotted near the old railroad track along the South Side Road in St. John's, or one seen near a barn in the Goulds. Coyotes are here to stay and they aren't just a creature of the woods. In other jurisdictions they have demonstrated an ability to live in urban areas and that seems to be holding true in Newfoundland. Pets and small children may well be at increased risk of attack. Even though such violent coyote-human interactions are small in number, the potential is there for such an encounter to occur. Newfoundlanders may have to take precautions with small animals and their children that they didn't have to take before.

Provincial government officials may attempt to downplay the risk posed by coyotes, but giving their growing numbers, and their dispersion into highly populated areas of the province like the Northeast coast and the Avalon Peninsula, there may well be heightened chances of an attack occurring on a small child. At the least, people may have to start using sturdy garbage containers to store their refuse in to avoid baiting coyotes into coming near residences.

Maine's experience with coyotes has many parallels to the Newfoundland case. Wolves were extirpated from Maine in the early twentieth century, much like Newfoundland. This left only the red fox as the major canid predator in both jurisdictions.

By the mid-1970s coyotes were becoming established in Maine and were spread throughout the state by the mid-1980s. As coyotes became more abundant in Maine, Hilton's research shows that there was increasing political pressure for the state government to impose "strong predator control measures." A vocal portion of the public even called for the eradication of the coyote. While Maine has taken steps, such as having wardens snare coyotes, the species has not been eliminated, but nor has there been the catastrophic loss of wildlife and agricultural livestock that was widely feared. However, it is worth noting that Maine is home to white-tailed deer and lacks the caribou found in Newfoundland.

It is also worth remembering that the example of the Gaspesie' caribou herd in Quebec indicates that the when coyote predation was lessened through intensified trapping, caribou calf survival did increase. This may lend support to the notion that heightened action is need to protect Newfoundland's caribou from the coyote's relentless jaws.

Biologist Gerry Parker concludes that "the most practical and effective program of coyote management in the northeast would appear to be one which encourages public use (trapping, hunting) and control, when necessary, through government extension services which encourage farmers and ranchers to solve their own problems."

It seems that this is the model currently being followed by the provincial government of Newfoundland. Clearly, Parker's research is highly regarded by biologists and wildlife managers. However, we must remember that Parker's work severely underestimated the ability of coyotes to prey on moose and caribou in Newfoundland during winter. Therefore, we must question the government's adoption of his policy recommendations given the flaw in his theory regarding coyote-big game interactions.

Is the provincial government not being proactive enough in dealing with the coyote in Newfoundland? Many sheep farmers, trappers, outfitters and hunters think so. Certainly, given the overall lack of research from other jurisdictions regarding coyote-caribou interfacing, it is disturbing that more is not being done to understand the coyote's full impact on Newfoundland's caribou herds. Given the cultural, social and economic important of caribou to Newfoundlanders, the provincial government must take a stronger role in fully understanding the impact of coyote predation on caribou.

... CHAPTER TEN ...

Conclusion

It seems the coyote is here to stay in Newfoundland. Hunters, outfitters, trappers, sheep farmers, big game managers and scientists all have to learn to live with the "barking dog."

Just as it has throughout all of its range expansion, the eastern coyote has generated controversy, fear and loathing. As Parker says:

> The arrival of the coyote in the northeast would have been
> heralded with far less fanfare, and certainly less
> controversy, if deer had not declined concurrent with
> coyote colonization and population increase.

Similarly, in Newfoundland, the coyote's arrival might not be generating such concern if it did not kill caribou, moose and sheep.

As Shane Mahoney points out, even if the main reason the current decline in Newfoundland's caribou is over-browsing of the range (i.e. food depletion) there will inevitably be some people who will blame the coyote. Certainly, a crash in the caribou herds at the same time that the coyote was expanding its range and densities does seem more than coincidental. But given the lack of research from other jurisdictions on caribou-coyote interactions the current and future state of caribou herds in Newfoundland is worrisome. The provincial government seems to be taking a very relaxed approach to the coyote that has many interest groups howling mad. For example, why are outfitters feeling totally ignored by government's management of coyotes? One would think that given the importance of the outfitting industry to the provincial economy that more would be done to carefully protect and manage big game herds. If over-browsing of the range is part of the reason for the caribou crash then that in itself hints at poor management. Clearly, the provincial government must take a proactive approach in managing the coyote, particularly in regards its interactions with other economically important species such as caribou.

Because it is such an effective predator, throughout much of its range expansion, the coyote has been subjected to open warfare of bounties, poison, traps, and guns. Nothing has been effective in eradicating them and the ease with which coyotes can cover large tracts of territory means that vacant sections of range do not stay empty long.

Parker writes that part of the "problem" generated by the coyote is that many humans grew up in the "unnatural predator free system" caused by the extirpation of the wolf throughout much of northeastern North America. Certainly, this was the case in Newfoundland. Historically, there was a large dog predator on the island who co-existed with the native caribou and the earliest people, such as the Beothuks.

But the wolf has been extinct on Newfoundland since about 1922. The primary predators on the island were the quiet black bear, which hibernated for the winter, the secretive lynx, and the cunning red fox.

The absence of the wolf meant that there was a lack of large predators which was filled in the mid-1980s when the coyotes' presence on the island was confirmed. Parker concludes that the coyote has "assumed the role of the dominant large predator in the forest ecosystem." Some wonder if coyotes are only consuming their share of what their cousin, the wolf, used to consume? That is, the wolf existed in Newfoundland historically, so isn't there room for its cousin the coyote?

It is important to note that the Newfoundland ecosystem is special because of the lack of prey species. Other jurisdictions have many types of prey, such as deer, skunks, racoons and porcupines. Newfoundland has a relative scarcity of prey and this may factor into the coyote's apparent reliance on caribou.

Shane Mahoney explains that, in the recent past, due to the absence of the wolf or a similar dog predator, big game managers have been able to set higher harvest levels for hunters. Obviously, the high predation rates of coyotes on caribou calves, and the predation on adult caribou, means that this will have to be factored into the establishment of hunting quotas. With calf recruitment low, many caribou herds are in a state of decline. Will there be steep reductions in hunting quotas for both residents and outfitters' clients? What will be the impact of such a move on outfitters, their guides and the economy? If the caribou herds drop very low, it will be years and years before they come back and a vital aspect of our culture and economy will be hurt. The people of this province deserve answers and transparency from wildlife managers and government. It seems astounding that there isn't more being done to understand the relationship between caribou and coyote.

Research in Maine by Henry Hilton offers some soothing conclusions about coyote interactions with other species. He writes that

after more than 15 years of coyote population growth leading to relatively stable population densities throughout the state, coyote predation has not caused the catastrophic damage to wildlife or agricultural interests that had once been widely feared.

Will the same happen in Newfoundland as coyotes reach stable population densities over the coming years? Or, will caribou and other species be decimated by the relentless, vicious jaws of the little wolf? Much remains to be learned and many questions are at this point unanswered. The provincial government must take a lead role in analyzing the issue and making results known to the public. Recently, the premier has gotten personally involved in the issue of salmon poaching. Perhaps he and his ministers need to become as focused on the island's caribou herds vis-a-vis coyote.

Outfitters such as Wayne Holloway and Eric Patey are concerned that caribou are headed for near extinction. "My prediction is we will see our woodland caribou placed on the threatened species list," Holloway says. Surprisingly, Shane Mahoney does not disagree with Holloway. "Since 1922 there's been no predation of adult caribou in the winter, we desperately need to get a handle on it...We will have far more coyotes than we ever had wolves. The territorial nature of wolf packs means they will self-limit, coyotes won't do that."

Retired wildlife officer and environmental consultant Bruce Porter agrees with Mahoney. "There's more coyotes than the ones you see. Their tracks are everywhere in the snow."

Mahoney also says that coyotes will feed on anything, will live close to people and they will not fall if the caribou population falls, but they will be there in bigger numbers to keep caribou down. His speculation is similar to the situation with white-tailed deer in Nova Scotia. According to that province's director of wildlife, Barry Sabean, "coyotes made the deer population crash faster, drove it lower and slowed the recovery of deer."

John Blake is a hunter and a wildlife manager. He says that "the bottom line is that we have a new predator on the block filling the void left by wolves in 1922. The interaction will play itself out largely to our ignorance."

Obviously, Newfoundland is in a tight financial situation. Many government departments and issues are competing for scarce funds. But given the very high participation rates in hunting in this province, it seems negligent of the government to essentially sit on its haunches and quietly watch the coyote story play itself out.

Field research conducted for this book found that Newfoundland's wildlife division is in tatters. Consecutive provincial governments have underfunded the division and cut it into pieces. It has been effectively beheaded and dismembered. For example, Building 810 in Pleasantville once housed the headquarters of the division. Now it is home to a handful of biologists and lab technicians. Many of its offices are vacant and the dusty corridors have a hollow echo as a visitor passes through. Other wildlife managers are scattered about in offices in Corner Brook and Pasadena. Trying to track down information on coyotes was sometimes very hard and forced me to make several phone calls to various offices. It seems that wildlife management and research in this province is in disarray and it needs attention from the provincial government. Too little is being done to fully comprehend the impact coyotes are having on the island's ecosystem. Hasn't the cod crisis taught us that we can't afford to ignore the cries of outfitters, trappers, sheep farmers and hunters?

Similarly, big game biologists also have much to learn about the coyote such as home range size, litter size, and actual impact on big game stocks, while big game managers must factor coyote predation into formulas when establishing moose and caribou hunting quotas. The provincial government must begin to funnel more resources into the science and management of wildlife in general, and coyotes in particular.

The recently created coyote license and ten month shooting season seems to be an example of inefficient and illogical government policy, however, well-intentioned it might be.

Shooting coyotes requires specialized, costly rifles and honed marksmanship skills. Sadly, many young hunters lack the latter. For example, several years ago the provincial government cancelled mandatory shooting tests for new big game hunters. That is, novice moose, caribou and black bear hunters are permitted to go in the field without having to prove that they can use a high-powered rifle accurately. Now, hunters are similarly being permitted to purchase a coyote license without any government employee being responsible to ensure that the hunter can hit the broad side of a barn with a rifle capable of killing a person a mile away. Is this responsible and effective wildlife management? Imagine the outcry if the government did away with the road test for new drivers. Yet, no one seems concerned that novice hunters can buy a license and take a high-powered rifle without proving their proficiency with that firearm. Many hunters are entering the field with rifles that they have never fired before. This may lead to increased chances of some one being seriously hurt or killed, and of wildlife being maimed, crippled or wounded and going away to suffer a slow death. If government is really serious about a shooting season for coyotes, it must re-introduce mandatory marksmanship testing for coyote hunters, and for all species hunted with high-powered rifles. To do otherwise is gross negligence.

Concomitant with legislated shooting tests, government should introduce training programs for those hunters wishing to specifically target coyotes. Coyotes are a new, distinct species on the island and hunters have no background in hunting them. The species requires specialized skills not required of gunners used to pursuing moose and caribou. Government has to put its money where its mouth is. If the coyote license is to be anything more than a politically expedient white-wash, government must begin to offer training programs.

Even though big game hunting is primarily recreational in Newfoundland today, most hunters still want meat for the winter. What are the ethics of pursuing and shooting wild dogs for sport? Such a hunt may leave the province open to complaints from the animal rights movement. At least trappers take fur-bearers for a sound economic reason. But having hunters with high-powered rifles shoot coyotes for pure sport seems rather cruel and wasteful. This situation of a coyote sport hunt is vastly different from that of the sheep farmer who shoots a marauding coyote.

Sadly, many of Newfoundland's sheep farmers feel alienated by government's coyote policies. A large number of sheep farmers have left the industry and those that have stayed have had to adopt measures to survive. Again, given the economic relevance of sheep farming to the provincial economy, and given the potential market for local lamb, it seems that the provincial government is clearly failing sheep farmers. This is not a criticism of specific government employees who no doubt are working hard for sheep farmers, but is rather a critique of government's over-all position. Sheep farmers I spoke with feel they have been left high and dry by government. Sheep producers feel that their very way of life is threatened and that no-one in authority cares.

There is unrealized economic potential in sheep farming and much of Newfoundland's countryside is suitable for such industry. And, Newfoundland has a long history of sheep farming that must be nourished as an important part of our history and culture. The threat posed by coyotes is very real and pressing and government must become more proactive in its treatment of the matter. Our sheep producers deserve no less.

Another important issue related to the coyote has to do with rabies. According to province's chief vet, Dr. Hugh Whitney, coyotes do not appear to be an effective carrier of the rabies virus. Although he is aware that the state of Texas recently had a big problem with coyote rabies and spent large sums of money dealing with it. Of course, Newfoundland has just come through the extensive rabies eradication program which focused mainly on red foxes. He says that throughout the program the vast majority of animals tested were red fox, followed by lynx and coyotes. None of the lynx or coyote tested positive for rabies.

Whitney explains that Newfoundland is home to fox rabies and we may have cases where a rabid fox bites another animal such as a cat, dog, or coyote and that animal then becomes infected. However, to date Whitney is not aware of any cases of rabid coyotes, although he cautions that it could possibly happen.

The chief vet also explains that coyotes may actually help reduce fox rabies on the island by helping to drive the red fox population down. Red fox are the more effective carriers of the rabies virus on the island. In the eastern Canadian experience foxes seem to live at higher densities than coyotes and this may help further the spread of rabies amongst fox.

Was the coyote introduced to the island of Newfoundland as some people believe? Or is the coyote's invasion of Newfoundland a remarkable testament to a truly resilient, highly adaptable, incredibly smart canid?

The very fact that many responsible citizens from various sectors like outfitting, trapping and hunting believe or suspect that government may have secretly introduced the coyote to replace the extinct wolf suggests that many people lack trust in the machinery of government. Again, this is not to say that bureaucrats are not trustworthy, but that the overall policies and government process regarding coyotes in particular and wildlife in general leave many stake-holders feeling alienated and ignored. Government needs to take remedial action to enhance transparency and accountability to tax-payers.

The reality is that Newfoundland has a long history of exploiting natural resources and an equally lengthy record of poor resource management. However it got here, the fact is that coyotes are here now, and here to stay. We need to know as much about this remarkably efficient predator as we can. The onus is on the provincial government to better understand the species, and its impact on other wildlife such as caribou, and to better educate the public about the rhetoric and reality of threats posed by coyotes on the island of Newfoundland.

Coyote crossing Little Grand Lake
- courtesy John Neville

Coyote near Granite Lake
- courtesy B. Sellars,
Nf & Lab Hydro Co.

Coyote tracks with sunglasses for size   - courtesy John Neville

Coyote near TCH by Port Blandford June 2004. - courtesy G. Tuff

John Blake with coyote - courtesy John Blake

G. Jennings Photo

George Jennings with first coyotes trapped in Newfoundland, 1986 near Corner Brook.

Llama belonging to Sandy and Laurie Ruby to protect sheep.   - D. McGrath

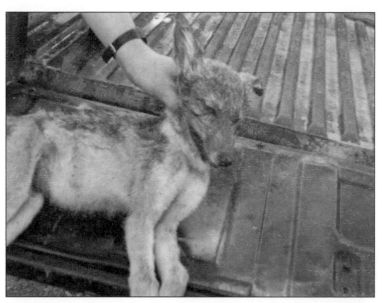

First confirmed coyote in Newfoundland. Hit by truck on Camp 10 Road near Deer Lake. Thought to be red fox, confirmed to be coyote pup. First confirmed coyote killed in Newfoundland.   - G. Jennings Photo

Maremma, guard dog owned by Wade Alley to protect sheep — P. Dunphy

Coyote near Granite Lake during winter 2004. — B. Sellars, NF Hydro Co.

Coyote on sea ice in the Northumberland Strait near PEI.
- Photo courtesy R. Dibblee.

John Blake and Wayne Barney in Serpentine Valley, Winter 2004 with coyotes called out and shot. - Courtesy John Blake

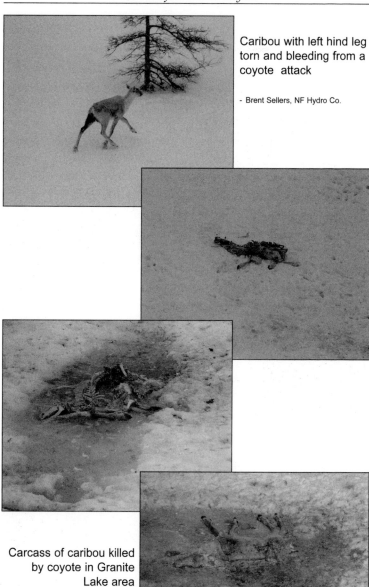

Caribou with left hind leg
torn and bleeding from a
coyote attack

- Brent Sellers, NF Hydro Co.

Carcass of caribou killed
by coyote in Granite
Lake area

- Brent Sellers, NF Hydro Co.

References

Acorn, Robert and Michael Dorrance. *Coyote Predation of Livestock.* 1998. Published by Alberta Agriculutre, Food and Rural Development.

Acorn, Robert and Michael Dorrance. *Methods of Investigating Predation of Livestock.* 1990. Published by Alberta Agriculutre, Food and Rural Development.

Chambers, Robert. "Reproduction of Coyotes in their Northeastern Range." Pages 39 - 52 *Ecology and Management of the Eastern Coyote* edited by Arnold Boer, 1991.

Connolly, Guy. "Coyote Damage to Livestock and Other Resources." Pages 161 - 169 *Ecology and Management of the Eastern Coyote* edited by Arnold Boer, 1991.

Coyote: General Biology, Management and Harvest Strategies. 2004. Information Pamphlet produced by John Blake, Inland Fish and Wildlife Division, Government of Newfoundland.

Dorrance, Michael. "Coyote Control in Alberta." Pages 171 - 182 *Ecology and Management of the Eastern Coyote* edited by Arnold Boer, 1991.

Harrison, Daniel. "Social Ecology of Coyotes..." Pages 53 - 72 *Ecology and Management of the Eastern Coyote* edited by Arnold Boer, 1991.

Hilton, Henry. "Coyotes in Maine: A Case Study." Pages 183 - 194 *Ecology and Management of the Eastern Coyote* edited by Arnold Boer, 1991.

Lavigne, Gerald. "Sex/Age Composition and Physical Condition of Deer Killed by Coyotes During Winter in Maine." Pages 141 - 159 *Ecology and Management of the Eastern Coyote* edited by Arnold Boer, 1991.

Litvaitis, John. "Niche Relations Between Coyotes and Sympatric Carnivores." Pages 73 - 85 *Ecology and Management of the Eastern Coyote* edited by Arnold Boer, 1991.

McNab, Leslie. "Invasion of the Trickster." Pp. 60 - 65 in *The Downhomer* July 2004.

Moore, Gary and Gerry Parker. "Colonization by the Eastern Coyote." Pages 23 - 37 *Ecology and Management of the Eastern Coyote* edited by Arnold Boer, 1991.

National Trapper's Association Handbook. www.nationaltrappersassociation.com.

Pekins, Peter. "Winter Diet and Bioenergetics of Eastern Coyotes." Pages 87 - 99 *Ecology and Management of the Eastern Coyote* edited by Arnold Boer, 1991.

Parker, Gerry. *Eastern Coyote: The Story of Its Success.* 1995, Nimbus Publishing, Halifax.

Wayne, Robert and Niles Lehman. "Mitochondrial DNA Analysis of the Eastern Coyote." Pages 9 - 22 *Ecology and Management of the Eastern Coyote* edited by Arnold Boer, 1991.

First-person interviews were primarily conducted during May and June of 2004. Subjects in the greater St. John's area were interviewed in person, while subjects living outside metro-St. John's were interviewed over the telephone.

Darrin McGrath is a free-lance writer and author. He has written for a variety of newspapers and magazines including: **The Halifax Chronicle Herald** where he has a regular column; **Outdoor Canada; Explore; The Navigator; The Downhomer and The Telegram. The Newfoundland Coyote** is his fourth book. His most recent work was the best-selling **Hitching a Ride: the Unsolved Murder of Dana Bradley.**

Darrin lives in St. John's with his wife Ann and their three dogs.

- photo Joe Gibbons